WHO'S AFRAID OF HUMAN CLONING?

WHO'S AFRAID OF HUMAN CLONING?

Gregory E. Pence

ROWMAN & LITTLEFIELD PUBLISHERS, INC.
Lanham • Boulder • New York • Oxford

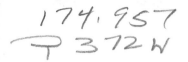

ROWMAN & LITTLEFIELD PUBLISHERS, INC.

Published in the United States of America
by Rowman & Littlefield Publishers, Inc.
4720 Boston Way, Lanham, Maryland 20706

12 Hid's Copse Road
Cummor Hill, Oxford OX2 9JJ, England

British Library Cataloguing in Publication Information Available

Library of Congress Cataloging-in-Publication Data

Pence, Gregory E.
    Who's afraid of human cloning? / Gregory E. Pence.
        p.   cm.
    Includes bibliographical references and index.
    ISBN 0-8476-8781-3 (cloth : alk. paper).—ISBN 0-8476-8782-1
(pbk. : alk. paper)
    1. Cloning—Moral and ethical aspects.   2. Human genetics—Moral
and ethical aspects.   3. Human reproductive technology—Moral and
ethical aspects.   I. Title.
    QH442.2.P46   1997
    174'.957—dc21   97-38513
                                                             CIP

ISBN 0-8476-8781-3 (cloth : alk. paper)
ISBN 0-8476-8782-1 (pbk. : alk. paper)

Printed in the United States of America

⊗ ™ The paper used in this publication meets the minimum requirements of
American National Standard for Information Sciences—Permanence of Paper for
Printed Library Materials, ANSI Z39.48–1984.

Within the last two years, bioethicists and philosophers David James of Old Dominion University and Benjamin Freedman of McGill University died suddenly at mid-life. They will be missed by the entire bioethics community.

Contents

CONTENTS

Preface

. . . . .

At the end of June 1997, I had the opportunity to meet the man who cloned Dolly, Ian Wilmut, and to discuss with him some of the issues discussed in this book. Although he does not agree with my position on human cloning, I have been enriched by meeting Ian Wilmut.

The opportunity to meet Dr. Wilmut came at the first major conference on human cloning after the announcement of Dolly, and it was organized in Washington, D.C.,—by French Anderson, Art Caplan, Alex Capron, and Craig Venter— to come just after the Report "Cloning Human Beings" was issued by the National Bioethics Advisory Commission (NBAC). Although I had already written over half of this book, this conference provided a great chance to improve it by meeting the leading people in the world in mammalian cloning, health law, genetic research, and bioethics. In addition, there were trademark attorneys and venture capitalists, experts in primate twinning, philosophy professors, and representatives of various religions. In short, it promised to be one of those extraordinarily exciting conferences in bioethics when the best people are pulled together to discuss a new development.

What was disappointing to me about the conference was that none of the speakers was willing to defend human cloning in any way. It seems that a consensus had developed among bioethicists, theologians, and scientists that this was a side not to be defended. In one of my only comments at this conference, I criticized the one-sidedness of the presentations and suggested that professionalism in bioethics meant that each side had to be defended with logic and passion, and that this had not been done. In reply, I was told that "perhaps a consensus has developed that human cloning should never be done and that is why only one side is represented."

But how can a "consensus" develop when the arguments are never made? When there is no discussion? No debate? I left disappointed.

There will be other books that discuss the events that led up to the cloning of Dolly and the reaction to that event. Journalists will rush to get books out, and there will be a lot of quotations in them from various people about what human cloning means.

In contrast, this book is unabashedly philosophical and one-sided. I am making the case for human cloning first because there is a terrible one-sidedness to the "discussion" so far and second, because I believe that my position is true.

Acknowledgments

· · · · ·

I owe many people for helping me to write this book and I wish to thank them here: Bob Blaylock, an editor at the *Birmingham News*, for getting me to write an Op-Ed on human cloning after the announcement of Dolly's birth; Rosemarie Tong and Lance Stell, for inviting me to Davidson College, where they and their students made several nice suggestions; microbiologist Susan Self, for inviting me to discuss human cloning in her honors seminar at UAB; the inaugural class of the executive masters' program in Health Administration, especially John Comstock, MD; the classes of 1999 and 2000 of the UAB Medical School. Sarah Moyers at McGraw-Hill, for allowing me to delay the 3rd edition of *Classic Cases in Medical Ethics*; James Pittman, the former Dean of the UAB Medical School, for advice, encouragement, and a constant stream of sources; Stephen Jay Gould, whose visit to UAB in May and whose brief discussion of human cloning with me at dinner helped me to understand the importance of conjoined twins in this debate; Harold Kincaid, G. Lynn Stephens, Florence Siegal, and the perceptive students in my senior seminar and bioethics classes in the first half of 1997. Philosopher Richard Mohr made helpful suggestions about gay issues and cloning.

I also wish to thank the librarians at the Lister Hill Library for Medical Sciences and at the Sterne Library, especially Diann Weatherly. The financial support of the UAB Medical School allowed me to attend the first conference on the ethics of mammalian cloning, for which I am grateful.

Biologist Thane Wibbels, Ona Marie Faye-Petersen of the Anatomical Pathology department and Jerry Thompson of Medical Genetics, all at UAB, reviewed various parts of the manuscript and answered my questions. UAB biologist Ann Cusick delayed packing for her trip to Alaska to make sure I understood embryology before she left. Mike Miller, while writing an honors thesis with me on human cloning, helped proofread this manuscript and gather sources for it.

I owe a huge debt to my next-door neighbor, James Rachels, who read the manuscript and made many valuable suggestions about organization, arguments, ethical theory, and strategy. I am deeply in debt for his help and encouragement. Another colleague, Mary Whall, carefully read the entire manuscript shortly before it went to press, performing in a supererogatory fashion and greatly improving the book. I am really in her debt, too. Robert Angus, who teaches various genetics courses at UAB, read key parts of the manuscript and corrected some of my mistakes (of course, what mistakes remain are mine alone). Sociologist Jay Hughes and an anonymous reviewer for Rowman & Littlefield were ideal, constructive, informative reviewers. My wife Pat has also supported this book from the get-go, read and commented on parts of it, and I also thank her for tolerating my absences during the first three months when I was so immersed in the details of cloning. My colleague G. Lynn Stephens generously helped proofread the manuscript on very short notice.

Christa Davis Acampora, in her time as Philosophy Editor at Rowman & Littlefield, and later her assistant, Robin Adler, were exceptionally helpful and enthusiastic about this project.

Mary Had a Little Lamb

Mary had a little lamb, its fleece was slightly gray
It didn't have a father, just some borrowed DNA.

It sort of had a mother, though the ovum was on loan,
It was not so much a lambkin as a little lamby clone.

And soon it had a fellow clone, and soon it had some more,
They followed her to school one day, all cramming through the door.

It made the children laugh and sing, the teachers found it droll,
There were too many lamby clones, for Mary to control.

No other could control the sheep, since the programs didn't vary
So the scientists resolved it all, by simply cloning Mary.

But now they feel quite sheepish, those scientists unwary,
One problem solved but what to do, with Mary, Mary, Mary.

Anonymous post on the Internet

From Dolly to Humans?

Buddhist scholar Donald Lopez foresees real problems for the theory of karma. Would the clone inherit the karma of the original person? And he wonders, "What did the sheep do in a previous life that resulted in its being cloned in this one?"[1]

·　·　·　·　·

It took about a second for the questions to begin. And another for the condemnations. Actually, there were not many questions, just condemnations, because thought stops when most people hear "cloning humans."

On February 24, 1997, every paper in the world carried a front-page story about a lamb named "Dolly" that had been cloned near Edinburgh, Scotland by Ian Wilmut. It was immediately apparent that Wilmut's cloning techniques might be applied to humans. Never in the history of modern science had the world seen such an instant, overwhelming condemnation of the application to humanity of a scientific breakthrough. Even physicians and scientists, who should have known better, joined the chorus of "Thou Shalt Not Clone Humans!". Wilmut himself insisted that no good reason existed to clone humans.

Thirty hours after the news of Dolly hit the streets, legislator John Marchi announced a bill to make human cloning illegal in New York State.[2] Conservative ministers and rabbis jumped on the same bandwagon. An official of the Catholic Bishops Conference of England and Wales urged banning human cloning because "each human being has a right to two biological parents."[3] National religion columnist Mike McManus immediately urged not just condemnation but a law to make human cloning illegal.[4] After ninety days of study, the National Bioethics Advisory Commission (NBAC) agreed, urging federal legislation to ban human cloning.[5] France's President Jacques Chirac pressed the Group of Seven for a similar ban. Chirac, who only a month before had held a computer's mouse for the first time, seemed not to understand what cloning might be, saying it "undermined the dignity of people by creating a desire to avoid death."[6]

Most bioethicists (or at least the ones who first speak in public) also did not enlighten the public much and immediately condemned human cloning. Invoking the oft-used club, they claimed that human cloning would slide us down the slippery slope. Richard McCormick, a Jesuit bioethicist at the University of Notre Dame, declared categorically that a person who would originate another from her genotype "is overwhelmingly self-centered."[7] Protestant bioethicist Alan Verhey declared that such origination would lead parents to think of their children as mere products.[8] At a hearing of the U.S. Senate committee, Boston University bioethicist and law professor George Annas testified, "I agree with President Clinton that we must 'resist the temptation to replicate ourselves' and that the use of federal funds for the cloning of human beings should be prohibited." Within days of the announcement and with no public debate having occurred, Annas confidently claimed that "there are no good or sufficient reasons to clone a human being."[9]

How could this be known, just days after Wilmut's announcement? Before anyone had time to discuss it? Before any arguments were put forth? We urge children to think critically about controversial moral issues before making up their minds—to examine both sides before forming opinions— but then President Clinton announces that human cloning should be banned only days after the news of Dolly, reflexively dismissing one of the biggest philosophical questions ever to emerge from biology.

These knee-jerk condemnations stem from fear and ignorance; they should not be mistaken for moral wisdom. To this author, the public's clonophobia was more interesting than the news about Dolly. And the numbers from the public were impressive: a Time/CNN poll conducted a few days after the announcement discovered that 93% of Americans disapproved of cloning humans.[10]

This book discusses the ethics of human cloning. It attempts to think through the questions that cloning humans raises at different levels, including cloning human embryos to help infertile couples, cloning embryos to study genetic disease, a married couple's origination of a baby through cloning to prevent a baby with a genetic disease, and the extent to which human cloning should be forbidden, permitted, limited, or even encouraged by public policy.

Ultimately this book discusses more than whether a few embryos are gestated to adulthood in a different way. Human cloning raises some important questions about our ability to choose wisely, about our view of human nature, about our capacities, about our faith in ourselves, and about the direction we chose for future humanity.

Up until now, there has been a tendency not to discuss such matters seriously and to leave them to futurists. But the future is here and we can no longer avoid thinking about cloning. In doing so, we want to avoid two mistakes. The first is the ostrich-in-the-sand approach, which pretends that such questions will go away. Many predict that some scientist somewhere in the world will originate a person through cloning. Better to think about such matters in a cool, calm way before they occur.

The second mistake is to look too close in front of us, and not to gaze far enough into the future. We must avoid only looking at the cases at hand. Most of bioethics focuses on such real cases and rightly so. Yet events such as Dolly's announcement prompt some of us to look farther off. Never looking really far ahead is like growing up in one neighborhood, always believing it's a great place to live, and never going anywhere else. If we don't take some risks and leave, we may never get a new perspective and live all our life in the wrong place.

Questions about the beginning and end of life are at the heart of bioethics and philosophy. They are the kinds of questions that philosophers have always thought about. Since the early 1970s, the new interdisciplinary field of bioethics (aka "medical ethics") has developed and attracted many philosophers. As science and new technology have pushed back the limits of medicine, bioethicists and others have thought about answers to new questions, such as the meaning of death and the morality of new ways of creating babies. And now we must address human cloning.

Questions about cloning are also philosophical in two other senses. First, most of the questions are not about scientific facts but about ethics, human nature, and public policy. Those are the questions of philosophical claims and analysis. Second, since the time of Socrates, philosophy, at its best, has always been about questioning assumptions. The status quo has decreed that it is unthinkable to clone a human being. To which philosophy responds: "Unthinkable? Let's think about that."

In our discussion, I will be fighting an uphill battle. A few days after the announcement of Dolly, President Clinton banned federal funds for human cloning, urging private biological firms to do the same. On March 4, he asked a little-known commission to make recommendations in 90 days about the ethics of human cloning. This National Bioethics Advisory Commission (NBAC) never seriously debated the merits of human cloning because it was obvious to its members what the correct conclusion should be. NBAC recommended a federal law to ban any attempt to create a human by cloning, presumably with stiff penalties to enforce it.[11] As such, it went well beyond the previous ban on federal funding of research on embryos and proposed a vast new area of federal intrusion into reproductive matters in what had hitherto been a hands-off area of private infertility clinics.

One member of the Commission, Alto Charo, explained that the Embryo Research Panel, on which she had served in 1994, had made quite reasonable recommendations about limited research on human embryos, but had done so too rationally and logically for the public's satisfaction. (The recommendations were rejected because of associated issues of "partial birth" abortions.) This time, she vowed, NBAC would aim its recommendations at that emotional level which was needed to win in politics.[12] Hence NBAC's unanimous condemnation; hence NBAC's explicit recognition of the American public's irrational fears of human cloning as a reason for a ban.

I disagree with her view. I don't think the only appeal that wins is to the emotions, that every position has to be dumbed-down, or that people must be patronized. People can read the best arguments and change their minds. One can appeal to evidence and logic. The point of making a recommendation in bioethics should not be to "win" by having the recommendations implemented, but to find the truth and to explain it.

It is true that ordinary people have had a very visceral disgust at the possibility of human cloning. Nevertheless, when the entire English-speaking world has such an instant reaction, that very reaction cries out for explanation.

I think that such an emotional reaction is both understandable and explainable. Most of us have seen many examples in science fiction where bad things happen when humans get cloned. Countless movies and novels have created a deep distrust of scientists. Over and over again in the print and visual media, fear-mongers have described only the most evil reasons why anyone would want to clone humans. So is it any surprise now that various pundits claim that Hitler-like dictators would use cloning to produce unthinking automatons or that children who were cloned would grow up as neurotic adults?

Originating a human through cloning would be different from human in vitro or "under glass" fertilization (IVF), where eggs are removed from a woman's ovaries and fertilized with sperm in a Petri dish. After a few days of growth, the resulting embryos are reinserted in the woman's womb, there to gestate nine months until birth. IVF is sexual reproduction; cloning is asexual reproduction. IVF matches two different sets of 23 chromosomes to create a unique individual with a new set of 46 chromosomes; cloning reproduces the 46 chromosomes of the donor.

On July 25, 1978, Louise Brown, the world's first IVF baby, was born to John and Lesley Brown in England. Today, she is a typical English young woman. Asked if her classmates ever ribbed her about her unusual origin, the pudgy Louise giggled and replied, "When kids want to tease me, they ask, 'How did you ever fit into a test tube?' "[13]

Today, we take IVF for granted, and during the 13 years between 1978 and 1990, about 24,000 American children were created by using this technique. But in the mid 1970s, critics argued that creating a human baby through IVF might harm the baby.

The NBAC members were hard-pressed to explain why they had proposed a law to ban human cloning. ("Concerns about the potential impact of cloning human beings . . . on public and private values and morale are quite real, but nonetheless difficult to articulate with precision" their *Report* concluded.[14]) We know that conservative Chicago bioethicist Leon Kass testified to NBAC that human cloning was "morally repugnant" and that NBAC should act "as if the future of humanity may lie in the balance."[15] Other commentators used equally dramatic appeals.

NBAC seems to have thought that recommendations in sync with the emotional reactions of most Americans would be accepted. Its *Report* repeatedly cited the strong "discomfort, even revulsion" of most Americans against human cloning, as if such revulsion was itself a moral argument against such cloning.[16] NBAC focused on guaranteeing safety to any child created by cloning, but seemed at a loss when confronted by the fact that no human child has a guarantee of safety when being conceived. Steven Holtzman, a NBAC commissioner and an officer of Millennium Pharmaceuticals, denied that the safety of the child was the real motivation for the proposed ban. "Safety," he said, "just can't be what's motivating us here."[17] Getting to the emotional issue of human cloning and the unanimous desire of the Commission to ban human cloning, he added, "there's a very human tendency to know what you want to to do but not be clear necessarily about why."

Philosophers and bioethicists are very suspicious about "knowing what you want to do" but not knowing "why." Especially on issues such as human cloning, our emotional responses may dictate what we want to do and may not be justified. Our feelings may just be the result of too much science fiction.

We cannot allow such unreflected emotion to rule our lives. There are dangers here, and some of us have been through all this before. Twenty years ago, a battalion of intellectuals adamantly opposed in vitro fertilization. Now some of the very same people are using some of the very same arguments to oppose another new form of reproduction.

Although ethics is partly based on emotion, it is also more than that. As reasonable creatures, humans want reasons why a certain act is judged to be wrong. If the balance of reasons favors one side over another, we know that the right side is the one with the better reasons. Ideally, the reasons for a view go together in a good structure that logicians like to call an "argument," i.e., a structure with premises that support a conclusion.

Now argument must start somewhere and that is in acceptance of premises. Our emotions, our worldviews, our experience, our savvy, and our reflection determine which premises we accept as starting-points. But my point here is that one must offer some reasons, some argument, and some premises to have an opinion that is considered reasonable. It is not enough to say that one finds homosexuality "repulsive" or inter-racial dating, "offensive." Such reactions are merely ignorant prejudices. It is only when we give reasons for our emotional reactions that we and others can know if we are feeling something justified.

The reason why this point is important is that so much of the condemnation of human cloning is not based on arguments. Most of it consists of a conditioned "yuck factor" or vague appeals to the conflict of human cloning with "our values." Left so vague, these alleged conflicts are not conflicts at all, but merely abstract versions of emotional aversions.

Emotions change, and for the better when confronted with evidence. The introduction of Artificial Insemination by Donor (AID) for infertile women was greeted by disgust and condemnation in the 1940s and now is regarded as socially acceptable. When amniocentesis became available in the 1960s to test for genetic conditions such as Down syndrome, some commentators wanted to ban the procedure because it would lead to "selecting" only kids who were perfect. (Here, as in cloning, the fears jumped light-years ahead of the real possibilities.) In the early years of AIDS, many people were irrationally afraid of getting HIV-infected from casual contact, donating blood, and mosquitoes. In the early 1970s, before most Americans understood it, in vitro fertilization had an approval rating of only 15%, whereas today its rating is over 70%.

So emotions without reasons do not constitute good moral arguments. Emotions can be justified, but they can also stem from our most primitive, prejudiced reactions. Our better selves will always demand a reasonable explanation of why we feel as we do.

Take one last example. Have people really thought through what it means to pass a federal law against human cloning? Years ago, Germany passed a federal law that banned all forms of genetic therapy. Now new forms of treating genetic disease are available but cannot be done in Germany until that law is overturned. Do Americans really want to make it a federal crime to help a married couple have babies free of genetic diseases?

Are we sure now that no case will arise in the next few years where a child could be so helped by creation through cloning?

On another matter, there is also a profoundly important role for women in this debate. In most discussions of cloning so far, this role has not been emphasized, which by itself shows just how silly the discussion has been. For a cloned human embryo to become a person, a real flesh-and-blood woman must gestate such an embryo for nine months and raise it as her child. More important, she must choose to do so.

In this book, we are going to contrast two large ways of looking at these issues. On one side lies defeat and fatalism: we can keep muddling along as we have been, accepting whatever lethal genes slam into our families and knock people down. Or we can begin to take a more assertive stance toward the future of humanity and begin to answer questions that have never been really asked before.

Our tendency is to let certain images control our thinking. Because themes in science fiction portray cloned humans as victims of injustice, or as perpetuators of injustice, we think that it would be unjust to clone humans. Because the Nazis had a stupid, misguided eugenics program, we cringe from any use of that word with public policy. Because of one controversial case of commercial surrogate motherhood, we ban all commercial surrogacy.

And yet we do not really accept fatalism. We intervene constantly to better humanity, from removing lead from paint, to speaking more words to babies to create more neural pathways, to testing embryos for inherited genetic disease. In suburban neighborhoods all across the world, parents spend their free hours taking their children to piano lessons, to gymnastic classes, and to computer camps. In all these activities, they act to better their children and future adults. Why not think about bettering these people in other ways?

Choice is scary but not choosing is to allow genetic fatalism to rule our lives. Personally, I opt for a different future, one involving neither state-controlled reproduction nor genetic fatalism, but informed people who choose to make their children as happy as possible. I also opt for a different attitude, one of open-mindedness to the possibility of originating a child by cloning.

Notes

1. *Newsweek*, 10 March 1997, 60.
2. Raymond Hernandez, "Ban on Human Cloning Urged in New York," *New York Times*, 26 February 1997.

3. Nicholas Coote, assistant general secretary of the Roman Catholic Bishops Conference of England and Wales, quoted by Ehsan Masood, "Cloning Technique 'Reveals Legal Loophole,' " *Nature* 38 (27 February 1997).

4. Mike McManus, *Birmingham News*, 2 March 1997, 4(C).

5. National Bioethics Advisory Commission (NBAC), *Cloning Human Beings: Report and Recommendations of the National Bioethics Advisory Commission*, Rockville, Md., June 1997, 109.

6. Denis Boulard, "Chirac Condemns Human Cloning, " *Associated Press*, 29 April 1997.

7. *Newsweek*, 10 March 1997, 60.

8. *Newsweek*, 10 March 1997, 60.

9. George Annas, "Scientific Discoveries and Cloning: Challenges for Public Policy," Senate Subcommittee on Public Health and Safety, Committee on Labor and Human Resources, 12 March 1997. Testimony posted on the Internet at this address: http://www-busph.bu.edu/Depts/LW/Clonetest.htm.

10. Katherine Seelye, "Clinton Bans Federal Money for Efforts to Clone Humans," *New York Times*, 5 March 1997.

11. NBAC, *Cloning Human Beings . . .* , 109.

12. "Little Known Panel Challenged to Make Quick Cloning Study," *New York Times*, 18 March 1997, 14 (B).

13. Carol Lawson, "Celebrated Birth Aside, Teen Has Typical Life," *New York Times*, 1 October 1993, A8.

14. NBAC, *Cloning Human Beings*, 92.

15. Leon Kass, "The Wisdom of Repugnance," *The New Republic*, 2 June 1997; Kass' oral testimony (2nd quote) in Eliot Marshall, "Mammalian Cloning Debate Heats Up," *Science* 275 (21 March 1997), 1733.

16. NBAC, *Cloning Human Beings*, 3.

17. Gina Kolata, "Analysis: Commission on Cloning—Ready-Made Controversy," *New York Times*, 9 June 1997.

Dolly's Importance and Promise

The results [of Wilmut and colleagues] are of profound significance.
Commentary in *Nature* by Colin Stewart, National Cancer Institute[1]

· · · · ·

This chapter discusses some of the scientific background of the accomplishment of the team at the Roslin Institute led by Ian Wilmut and Keith Campbell, funded by the British government and the PPL Therapeutics of Ron James.

What Ian Wilmut Did and Why It Was Important

Ian Wilmut is a regular guy. Compared to the big bucks that American researchers commonly seek (and to the arrogance of some about the size of their grant monies), the $60,000-a-year government employee is very modest. He didn't complain that he would probably make only about $25,000 for his discovery. Although not a believer in God himself, he believes in ethics. His wife is an elder in the Church of Scotland.

Wilmut had been prodded to try cloning a lamb by a story he had overheard in a bar ten years before. While attending a scientific meeting in Ireland, Wilmut heard colleagues describe how Danish scientist Steen Willadsen, working at Grenada Genetics in Texas, had created a lamb clone by (what is technically called) "enucleating" an egg (i.e., removing the nucleus from the egg) and fusing what was left with a cell from a growing preembryo. ("Preembryo" is short for "preimplantation embryo" in embryology.)

Willadsen is widely regarded as a genius who pushes the outer edge of the envelope of scientific research. "The role of experimental science is to break the so-called laws of nature," he said in an interview in 1997.[2] He also said that he spends his summer "checking fences," which meant that he looks for holes in the stone wall of supposed scientific fact, seeking ways to get past what others had mistakenly viewed as solid and substantial.

In his first three attempts, Willadsen had actually succeeded in producing a live lamb, and in other (unpublished) work had done far better, having cloned cells from embryos that had 120-cells, in contrast to the usual 8-celled embryo.[3] In other words, the 120-cell embryo had already started to create differentiated cells, so Willadsen's alleged results here of having created new, undifferentiated embryos from these cells, if true, were the base upon which Wilmut worked. When Wilmut heard rumors of Willadsen's work, it was enough to draw him into similar work for the next decade.

The rest of the world went in a different direction, pursuing basic research in the new field of molecular biology rather than (what was thought to be) outdated embryology.[4] Part of the reason was a stunning example of the disastrous harm of fraud in science. The consensus that adults could not be cloned from differentiated cells occurred prematurely in part because of the uncovering of one of the most famous cases of fraud in recent history.

In 1981, Dr. Karl Illmensee and Dr. Peter Hoppe, of the University of Geneva and the Jackson Laboratory at Bar Harbor, Me., submitted papers to scientific journals, saying that they had proved that nuclei of adult mouse cells could be inserted in mouse eggs to produce adult mice.[5] The news was spectacular, like the news of Dolly's cloning, and a picture of the three cloned mice graced the cover of *Science* magazine. The two researchers were greeted as world heros for awhile, until researchers were unable to duplicate their results. Then doubt set in, and finally it was discovered that the dark colored spots on the white mice were fake, having been clumsily drawn on the coats with a black marking pen by Dr. Illmensee.[6]

The resulting despair was great. Some top researchers decided that embryology was a dead-end field after a devastating review article concluded that cloning adult mammals was impossible from cells that had specialized or "differentiated." Such differentiation occurred early in the life of the embryo. The severity of the reaction undoubtedly stemmed in part from the harshness of the reaction against the fraud.

It is interesting ,that in the creation of baby Louise Brown by in vitro fertilization, physiologist Robert Edwards had to disprove, like Ian Wilmut, several "facts." Edwards discovered that successful conception requires delicate, precise timing of interacting hormones. It had been thought that gonadotrophic hormones could not make a mammalian ovary release eggs, but Edwards learned how to do so. Human sperm must also be capacitated for conception; that is, chemicals that inhibit penetration of an egg must be removed from the head of a sperm. Most scientists had believed that capacitation required exposure to uterine secretions, and that this could only be done inside the uterus, but Edwards also proved that false.

Background on Cloning

To understand Wilmut's achievement and what it means for human reproduction, it is also important to understand some background. "Cloning" is an ambiguous term, even in science, and may refer to molecular cloning, cellular cloning, embryo twinning, and nuclear somatic transfer (NST). Only the latter is what occurred in Dolly and the latter is what most people care about regarding human cloning. In molecular cloning, strings of DNA containing genes are duplicated in a host bacterium. In cellular cloning, copies of a cell are made, resulting in what is called a "cell line," a very repeatable procedure where identical copies of the original cell can be grown indefinitely. In embryo twinning, an embryo that has already been formed sexually is split into two identical halves. Theoretically, this process could continue indefinitely, but in practice, there is a limited number of times an embryo can be twinned and re-twinned.

Finally, there is the process of taking the nucleus of an adult cell and implanting it in an egg cell where the nucleus has been removed. A variant of this process called "fusion" (which was actually done to produce Dolly) is to put the donor cells next to an enucleated egg and "fuse" the two with a tiny electric current. A blastocyst, a preembryo of about a hundred cells or less, starts to develop because the pulse that produces fusion also activates egg development. In fusion, mitochondria from both the donor and the egg recipient are mixed, whereas in strict transfer of a nucleus, mitochondria are only present in the enucleated egg.[7]

In 1993, American journalism and many bioethicists were breathless for a few days when the *New York Times*, on the first page of its widely read Sunday edition, reported that scientist Jerry N. Hall had cloned 17 human embryos into 48 as a way of increasing the supply of embryos in fertility clinics.[8] But these scientists had merely twinned an undifferentiated embryo, and had not "cloned" it in the ordinary meaning of nuclear somatic transfer. Instead, Dr. Hall had introduced an electric spark that causes unspecialized embryos of a few cells to twin, a process that had been done for years in the livestock industry. Although this was a useful technique to help infertile couples who could not produce enough embryos, it was neither a breakthrough nor startling. (More on the overreaction of bioethicists to this event in the next chapter.)

What Ian Wilmut did with an adult lamb was to create the lamb Dolly from differentiated, specialized cells of her adult ancestor. That this could not be done was considered such an ironclad fact that almost all promising biologists had gone into molecular biology and fled old-fashioned embryology (the American ban on federal funding for research on human embryos—discussed in Chapter 7—also fueled this exodus).

Before 1997, Ian Wilmut had spent years trying to clone one lamb. What Wilmut and his team did was to time the cycle of interphase of the cell cycle of the donor cell and recipient egg, such that the two were not out of sync. When fusion was previously attempted in this rapidly-changing, cellular environment, mitosis messed up, producing broken chromosomes and, hence, mutants and defective embryos. Wilmut starved eggs cells to make them dormant—stopping all the activity—and then fused the donor nucleus with the enucleated egg. In this way, they finally got a good fit, and at the right time.

Perhaps of equal importance was the fact that Wilmut and his colleague, Keith Campbell, were able to use nuclei from cells grown in a lab culture. A preembryo only has a few available cells to use for research, but the supply of cells in a cell line is endless. According to a summary in *Science News*, many labs had for years been trying to clone lambs and other livestock using cell lines, and the Roslin Institute finally did it with its lamb cell line.[9]

At a 1997 conference on mammalian cloning, Wilmut himself stressed several points. First, his present techniques are very inefficient: he started with 277 sheep embryos and got only one live lamb. He will not know until others repeat his work whether he was very lucky or very unlucky with his initial results: the real odds of a live lamb this way could be either 1/100 or 1/10,000. Second, his techniques have caused some examples of "large offspring syndrome," meaning that babies born are sometimes monstrously large and can't be born by normal means. This syndrome could be caused by nuclear transfer or by something in the uterine environment or by something in the cell culture. Third, the lambs that died shortly after birth were chromosomally normal. Fourth, he thinks we should avoid thinking of cells as "differentiated" or "undifferentiated," but instead think of differentiation as a continuum, where a cell's points on it are not forever fixed.

He believes that his techniques offer great promise for humans. Indeed, many scientists think that his real achievement may not be in cloning but in allowing us to understand and control cellular differentiation, to derive undifferentiated cells from differentiated cells, to understand how cells age, and to treat diseases caused by mitochondrial DNA. There is also the possibility of cell-based therapy with fusion of a nucleus to an egg for some diseases. More radically, and to avoid the ethical objections of a few people about using embryos created for research, there is the possibility of using de-differentiated cells as they are, without fusing them to an egg, to create an embryo for reproductive experiments.

He also thinks his techniques will be used in biotechnology to accomplish gene targeting, the insertion of a specific human gene in every cell of a lamb's body. It is important to stress that Wilmut did not start out intending to be the first person to create a live-born mammal by nuclear somatic transfer from an adult, differentiated cell. Instead he wanted to create a reliable method of inserting a new gene into every cell of a mammal. The old way of trying to do so rested on random chance, with very low probabilities of success.

(Incidentally, English animal rights activists have long opposed such research on animals. They had burned two labs at the Roslin Institute six years before Dolly was born.[10])

Five months after the announcement of Dolly's cloning, the team at the Roslin Institute announced that they had accomplished their goal.[11] The first lamb with a human gene was called Polly because she was a Poll Dorset Sheep. Lee Silver, a mouse molecular geneticist at Princeton University said of this achievement, "After Dolly, everyone would have predicted this, but they were saying it would happen in 5 or 10 years."[12]

The most exciting prospect here is to modify a sheep CTFR gene to create a model of cystic fibrosis (CF) for gene therapy. (In a classy move, Wilmut sold the first wool shorn from Dolly to raise money for the care and treatment of kids with CF.) Finally, Wilmut's techniques should help create genetically-altered organs of pigs, such that these new organs would have less chance of rejection in transplantation into dying humans. Similar modifications in livestock could also combat a disease called scrapie that affects the livestock industry.

What Cloning Can and Cannot Physically Reproduce

A person originated by cloning would not be an exact copy of an adult human being in many senses. At the molecular level there would be differences, even though the gene structure would be very similar. Atoms combine to forms molecules that in turn compose enzymes and proteins. At that point, two embryos starting out the same atomically reveal minor variations. For example, according to a chemist who thinks about such things, the probability that any two hemoglobin molecules in a human body are the same is "close to zero."[13] The probability of any two cloned human bodies being identical down to their last cell is virtually zero because the probability of any two things being exactly identical goes way down as the complexity goes up. When the jump is made from molecules to cells, complexity jumps exponentially because molecules can be combined in thousands upon thousands of ways to form cells.

As *New York Times* science writer George Johnson explained:

Even cloned cells, with identical sets of genes, vary somewhat in shape or coloration. The variations are so subtle they can usually be ignored. But when cells are combined to form organisms, the differences become overwhelming. A threshold is crossed and individuality is born.

Two genetically identical twins inside a womb will unfold in slightly different ways. The shape of the kidneys or the curve of the skull won't be quite the same. The differences are small enough that an organ from one twin can probably be transplanted into the other. But with the organs called brains the differences become profound. . . .

The precise layout of the cells, which neuron is connected to which, makes all the difference. Linked one with the other, through the junctions called synapses, neurons form the whorls of circuitry whose twists and turns make us who we are. In the reigning metaphor, the genome, the coils of DNA that carry the genetic information, can be thought of as a computer directing the assembly of the embryo. Back-of-the-envelope calculations show how much information a human genome contains and how much information is required to specify the trillions of connections in a single brain. The conclusion is inescapable: the problem of wiring up a brain is so complex that it is beyond the power of the genomic computer. The best the genes can do is indicate the rough layout of the wiring, the general shape of the brain. Neurons in this early stage, are thrown together more or less at random and then left to their own devices. After birth, experience makes and breaks connections, pruning the thicket into precise circuitry. From the very beginning, what's in the genes is different from what's in the brain. And the gulf continues to widen as the brain matures.

Even genetically identical twins, natural clones, are born with different neural tangles. Subtle variations in the way the connections were originally shaped together might make one twin particularly fascinated by twinkling lights, the other drawn to certain patterns of sounds.[14]

The brain, the most complicated human organ and the most essential to the continuity of our self, cannot be cloned or duplicated from a DNA blueprint. More importantly, the unique development of my brain that is the basis of my experiences can't be replicated in any sense by cloning. This should comfort some people who worry that individuality is threatened by cloning.

The New Genetic Age

Before going any further, a brief sketch of some facts of modern genetics will provide a basis for later discussion.

The basic unit of life is the cell, and in it the basic unit of heredity is the gene. The sum of all genetic information in the cells of an individual is called his or her "genome." How a genome is expressed, i.e., which characteristics an individual has, is called the "phenotype," the observable physical characteristics, and it is the result of the interaction of the genome and environment.

Genes are carried on chromosomes; a chromosome—which consists of DNA and other material—is a macromolecule composed of repeating nucleotides. Normally, the nucleus of each human cell contains 46 chromosomes. The germ, or sex, cells, however, have 23 chromosomes each; thus the union of sperm and egg provides the 46 (23 + 23) chromosomes for every new human being, and genes are inherited in pairs consisting of one gene from the father and one from the mother.

Genes consist of DNA, although not all DNA takes the form of genes. As a result of the work of James Watson, Francis Crick and others, it is now known that packed inside each of the 46 chromosomes is an enormously complicated strand of interwoven DNA: this is the famous double helix, strands of DNA (deoxyribonucleic acid), the nucleic acid responsible for transmitting hereditary characteristics. Each strand is composed of combinations of four chemical (nucleotide) bases in approximately 3 billion pairs (in each pair, one bit is on each side of the helix). The pattern of the four nucleotide bases in the 46 double helices is a person's genetic code.

If we think of a strand of these 3 billion base pairs as, say, North America, the Human Genome Project can be said to be providing a map of the territory and its major highways. Scientists believe that about 80,000 sequences of these billions of pairs of bases are genes, with the number of genes varying from chromosome to chromosome. On this emerging map of the human genome, the genes are the towns and cities, the destinations of most human travelers. This 15-year study, which began in October 1993 and is expected to cost $3 billion, is one of the most significant single projects in the history of science.

The largest gene, comparable in size to Los Angeles, is the gene for muscular dystrophy, composed of 2 million base pairs. The genes for globulin and insulin are like towns, with only about 1,000 base pairs each. At the beginning of the project, scientists already knew the location of most of the large cities and many of the smaller cities, but many of the towns remained to be found.

Because only 3% of DNA in humans actually codes for genes, most of the "land" of this map is vacant or "junk" DNA (not "trash" but "junk" as Sidney Brenner says, meaning that you don't throw out "junk" but keep it in case you later learn it is valuable). Right now, we have no idea what this 97% of DNA does for humans, if anything. At the start of the Genome Project, we knew about 2,000 genes in the human genome, and in 1997 we know about 50,000.[15]

Comparing macroscopic and microscopic levels is interesting. All the differences between any two humans is contained in only one-tenth of one percent of their DNA. Moreover, because of recent common ancestry in evolution, we share a very high percentage of genes with chimps.

Genetic diseases are inherited disorders. It is possible that most of us actually have genes for inherited disorders but are "heterozygous" for these disorders: that is, we have a dissimilar pair of genes for an inherited disease or trait. If a gene is "dominant," it will be expressed, or shown, in a heterozygote; if it is recessive, it will not be expressed in a heterozygote. However, even though heterozygotes may not show a trait themselves, they are carriers who can pass the gene for the disorder along to their offspring. If two parents who are heterozygous for a disorder both pass on the gene for the disorder to an offspring, that person will be "homozygous" for the disorder—will have an identical pair of genes. A disease or trait will always be expressed in a homozygote.

There are some exciting prospects on the horizon. Now that Wilmut and others have proved that cell differentiation is not an all-or-nothing process but a continuum and a process that can be reversed, hope abounds for treatment of neurological diseases. One researcher reported that neural growth factor (NGF) can turn off tumor-differentiated cells in 9 days. Obviously, such techniques could be used to create special neural cells from embryos to treat people with Parkinson's or Alzheimer's diseases or, even better, to change the neural cells inside these people to healthy, neural cells.[16]

As many as 15 million Americans may have moderate to severe genetic disease.[17] According to the Online Mendelian Inheritance in Man web-site maintained by the National Library of Medicine, there are over 6,678 Mendelian traits and disorders.[18] These are large figures; in fact, every family may include someone who is a potential victim of genetic disease or is susceptible to a disorder that may be linked to genetic causes, such as alcoholism, cancer, or coronary artery disease. Genetic diseases are estimated to account for over one-third of acute-care hospitalization of children under 18.

Although some diseases are caused by single-gene defects, such as cystic fibrosis and Huntington's, most diseases with genetic causes will be multifactorial in two senses. First, they may be caused by more than one gene (even single-gene diseases usually have more than one genetic variation). Second, they may only cause disease when they interact with specific factors in the environment. Thus the study of exactly how a disease is caused genetically is a complicated business. (More on this topic below, when I discuss the history of embryology and genetics.)

Genetic Contributions of the Host Egg

At the level of DNA, some DNA would come from the host egg, especially mitochondrial DNA. Mitochondria are the organelles (tiny organs) that fuel cellular reactions. They provide energy by metabolizing glucose, which in turn produces ATP and NADPH+, products which drive cellular reactions. It is thought that mitochondria, millions of years ago in evolution, may have once been bacteria. In mammals, mitochondria are inherited in the female cytoplasm (egg), which is separate from the nuclear genes from the female that are inherited in sexual reproduction.

An egg cell contains hundreds of mitochondria, which are randomly distributed among the new cells being produced by the growing embryo. Mitochondria carry genes, but not every cell has the same bit of DNA that is a gene. Because of aging of cells and environmental causes, the mitochondrial DNA can change over many years, which is thought to be a cause of some diseases such as Alzheimer's disease, Parkinson's disease, some forms of diabetes, and other more esoteric diseases. No one really knows how much mitochondrial DNA contributes to genetic diseases, in part because no one really knows how many diseases and disorders are solely caused by genes.

Both the nucleus inserted into the enucleated egg and the mitochondrial DNA have been affected by the life-time exposures of the adult to the environment. This is usually called the "environmental effects" on the genes.

Ian Wilmut recognized this point when he was asked whether he knew if the cells of Dolly would prematurely age. In other words, are these cells already 6 years old? Is Dolly only an (old) sheep in a (young) sheep's clothing? There was speculation that the udder cells of a ewe might be especially susceptible to environmental influences such as radiation and antibiotics, and hence, be different than the fresh cells of a baby lamb. When asked whether Dolly should be considered 7 months old, which is how long she had been alive at the first press conference, or 6 years old, as a genetic replica of a 6-year-old sheep, a reporter wrote, "Dr. Wilmut's clear blue eyes clouded for a moment. 'I can't answer that,' he said. 'We just don't know.' "[19]

Wilmut also did not know whether the donor DNA for Dolly came from mammary tissue or from a relatively undifferentiated mammary stem cell.[20] Stem cells are thought to be more malleable and capable of assuming new functions than other, more rigidly-defined cells. For example, stem cells are important in bone marrow transplantation.

Stephen Jay Gould doubts whether we should even regard Dolly's originating cell as an "adult" cell in the ordinary sense of the term:

Since the breasts of pregnant mammals enlarge substantially in late stages of pregnancy, some mammary cells, although technically adult, may remain unusually labile or even "embryo-like" and thus able to proliferate rapidly to produce new breast tissue at an appropriate stage of pregnancy. Consequently, we may be able to clone only from unusual adult cells with effectively embryonic potential, and not from any stray cheek cell, hair follicle, or drop of blood that happens to fall into the clutches of a mad Xeroxer.[21]

So the assumption that Dolly is an exact replica of anything is certainly false and there are at least two reasons why, even genetically, Dolly will not be the same as her ancestor. First, the host egg of the black-faced ewe, into which the nucleus of a white-faced lamb was put, had mitochondrial DNA that carries a few dozen genes. Second, we do not know whether this mitochondrial DNA shifted or changed over the lifetime of the ancestor ewe, such that Dolly inherited aged cells or some disease in these dozens of genes. No one will know until Dolly and other cloned lambs are closely observed over many years. Third, there is the possibility that Dolly's "adult" cells were some combination of aged, adult cells of the black-faced ewe and semi-embryonic cells, a combination that could vary with each cloning, such that there will be some variability in how close the next lamb clone from the same genotype will be to Dolly. Wilmut says that whether he truly accomplished nuclear transfer from an adult, differentiated cell can't even be proved definitively (since a stray stem cell could always be involved), but the proof can only accumulate as more and more successes are obtained.

Gould argues that identical human twins are more identical than cloned humans would be. Because Gould is known for paying attention to both heredity and the environment in their complex interactions, he is an appropriate person to emphasize that strict human identity not only requires an exact genotype inserted into a enucleated egg, but also the same inheritance of maternal cytoplasm (mitochondria), the same womb (and exposure to the same unusual events in the womb, e.g., alcohol consumption, falls), the same parents, and raising the children in the same geographical place and time in history. ("Does anyone believe that a clone of Beethoven would sit down one day to write a Tenth Symphony in the style of his early-nineteenth century forebear?") The case par excellence for this claim comes when Gould emphasizes that, "Eng and Chang, the original Siamese twins and the closest humans ever 'cloned' of all, developed distinct and divergent personalities. One became a morose alcoholic, the other remained a benign and cheerful man." Eng and Chang even had separate wives and alternated living in separate houses in North Carolina, where they fathered twenty-one children and lived to the age of 63.[22]

Gould's example is fascinating because it is a counter-example to the claim that cloned humans would be identical. (Note: the term "Siamese twins," originated by P.T. Barnum, is now considered offensive and "conjoined twins" is preferred.) "Every conjoined twin pair arises from a single zygote that did not quite succeed in staying together long enough to remain a singleton."[23] Conjoined twins originate from the same zygote, share the same womb, and share exactly the same environment for life (as one textbook of mammalian physiology states, "A clone is a group of genetically identical individuals; any pair of monozygotic twins is thus a clone of two."[24]) Yet conjoined twins originating from the same zygote and gestated in the same uterus sometimes manifest distinct personalities as adults.

Like Eng and Chang, the famous Tocci twins (a picture of whom was on the *Scientific American* cover of 1891), had different personalities. Like Eng Bunker, Giovanni Tocci drank beer in considerable quantities while Giacomo did not like beer and preferred mineral water. (The Tocci twins had separate livers and circulation; Eng and Chang shared a liver.) Giovanni was fond of sketching and introverted; Giacomo was the extrovert, a big talker, and also had a volatile personality (if he found some fault in Giovanni's sketch, he would kick the drawing off "his" knee). Similarly, the conjoined twins Abigail and Brittany Hensel, who share a common body below the neck, feel hungry and sleepy at different times, and sometimes get different grades on exams.[25]

Someone who thinks all differences in personality must have a genetic explanation will here point to the small differences that occur in how much of the X chromosome is shut off in each twin. Each member of female conjoined twins inherits one X chromosome from her mother and one from her dad. The phenomenon of X-chromosome inactivation occurs in early embryonic development when each cell randomly inactivates one of the two x-chromosomes: about half of the cells inactivate maternally-derived X and about half, paternally-derived X. Once a cell inactivates a particular X, that X remains inactivated in all cells descended from it via mitosis. Because the population of cells is not enormous at the time that chromosome inactivation occurs, the resulting distribution of cells inactivating the maternal X and those inactivating the paternal X is not always 50:50. For example, a female may express the maternal X in 70% of her cells and the parental X in only 30%.

Second, because of what is called "genetic imprinting" in autosomal genes, not all chromosomes are imprinted the same way, and not all parts of all chromosomes are always imprinted. Superimposed on chromosome inactivation is this variability of genetic imprinting.[26] Such small changes at such early stages of embryonic development, could account for later differences in "identical" conjoined twins.

Similar small changes in mitochondrial DNA, or changes in DNA from environmental effects in the adult ancestor of a cloned human, could create later differences in personalties or traits of two humans created from the same genotype. A final way to emphasize the same point is to emphasize that even with cows created from the same, cloned embryos, the model that geneticists use to predict similarity in milk production gives only a 70% match in taste and quantity.[27] Even identical calves do not have exactly the same uterine environments, and have slightly different postnatal environments.

In sum, both conjoined twins and two cloned persons would be very similar genetically, but still have small differences. Although the genetic contribution would create gross similarities, there would be enough differences that the two cloned people would seem to others as unlike as two conjoined twins or two "identical" adult twins. In both cases, there is some genetic background for such differences.

The Mistake of Genetic Determinism

The above points are instances of a deeper problem in understanding the role of genetics in producing humans. We live now in an age where people are sympathetic to (what is often called) "genetic determinism," the idea that the characteristics of a human are totally caused by his genes ("it's all in the genes"). The successes of the Human Genome Project have predictably fueled this attitude.

There is an ancient debate in biology going back to Aristotle about how much of human reproduction is fixed by something fixed and formal and how much of reproduction is fluid. Aristotle incorrectly thought that only males contributed the formal cause in their semen, and that such semen acted on mostly passive matter. "Preformationism," the idea that the development of the embryo is merely the unfolding of something present from the start in the seed, was found in human embryology even before Aristotle (e.g., in the Hippocratic corpus).[28] Preformationism dominated the medieval period, holding that humans were formed immediately at conception, such that embryogenesis was merely the expansion of a very small human (called a "homunculus"). According to this view, male ejaculation passed on a very small human; so Dalenpatius (1699) claimed to see homonucli in human sperm.[29]

Genes are not best understood just as simple mechanisms that cause traits in humans but as functions whose expression in individuals is mediated by what happens in the environment. The "expression" of a gene is not contained in itself, but is determined by how the genes of a particular individual interact with many other things.

The author of the scientific section of the NBAC *Report* put it this way:

Indeed, the great lesson of modern molecular genetics is the profound complexity of both gene-gene interactions and gene-environment interactions in the determination of whether a specific trait or characteristic is expressed. . . . recent scientific findings have revealed that a "one-gene-one-disease" approach is far too simplistic. Even in the relatively small list of genes currently associated with a specific disease, knowing the complete DNA sequence of the gene does not allow a scientist to predict if a given person will get the disease.[30]

We know, of course, that the environment plays a key role in the formation of individuals during childhood. The truth, however, is even more complex. Not only do the expressions of genes in creating a phenotype depend on the early child's environment, but also (and what is not so readily appreciated) is that the expression of genes even in embryonic development depend to an unknown degree on the uterine environment. Much of the expression of genes in the phenotype is not caused by preformatic, fixed templates (genes) that shape the passive matter of the embryo, but rather, the phenotype is caused by genes acting more like variables that can respond over a continuous range, depending on what happens around them.

To greatly oversimplify, we know that there are some, rare, powerful genetic traits such that if an individual has gene X, the trait is always eventually expressed. The gene for Huntington's disease is like this, being fully "penetrant" if one lives long enough. Other diseases and traits need a gene from both parents for the disease or condition to be expressed, e.g., cystic fibrosis or blue eyes.

More important to us, we do not know how much or how little the uterine environment influences genetic regulation (as opposed to genetic structure) of embryonic development in the first 100 days of human life. We postulate that many traits or characteristics of adults may require a certain set of genes from one parent mixed with another parent, and with this mixture developing in the right kind of environment. That is "another" kind of "genetic" causation. Again, we don't know all of what happens here. We know that the presence or absence of certain proteins or amino acids influences development of the brain. For example, in the genetic disease of phenylketonuria, the presence of excess phenylaline leads to an excess of phenylpyruvic acid that impairs development of the brain and results in mental retardation. But there may be all kinds of other causes of brain development and formation of personality of which we are now ignorant.

Finally, once the fetus has gestated and become a baby, a tremendous amount of shaping of later traits occurs within the first two years of life. Genes have shaped some ranges of response here, but what happens now depends on parents, nutrition, and all the other factors in a child's world.

In sum, the common view that it's "all in the genes" is false. But it is false in more complicated ways than at first appears.

Conclusion

Ian Wilmut's achievement was important, partly because he persisted in an area where others had given up and because he disproved putative facts in his field.

But his results should not be exaggerated. We cannot duplicate from genes our most important organ, our mind. Moreover, we cannot duplicate by cloning our own experiences and our memories, thus our unique identity is preserved.

Also, even a cloned human would not be a perfect copy of his ancestor. Small changes from mitochondrial DNA and environmental effects on the older DNA of the ancestor may be passed on to the new person. The exact nature of the identity will not be known for some years (e.g., until we see when Dolly dies.)

Finally, some of the misunderstandings of human cloning impute too large a role to genetic determinism, especially in its gene-centric or "preformationist" versions. Even embryogenesis may depend on what's in the uterine environment, such that two genetically identical embryos could turn out as very different babies based on what happened in the mother's uterus.

Notes

1. Colin Stewart, "Nuclear Transplantation: An Udder Way of Making Lambs," *Nature* 385, no. 6619 (27 February 1997), 769, 771.
2. Gina Kolata, "Iconoclastic Genius of Cloning," *New York Times*, 3 June 1997, B7, 12.
3. Gina Kolata, "Iconoclastic Genius," B12.
4. Michael Specter with Gina Kolata, "After Decades and Many Mistakes, Cloning Success," *New York Times*, 3 March 1997.
5. Michael Specter with Gina Kolata, "After Decades . . . ."
6. Michael Specter with Gina Kolata, "After Decades . . . ."
7. "The Science and Application of Cloning," National Bioethics Advisory Commission, *Cloning Human Beings: Report and Recommendations of the National Bioethics Advisory Commission*, Rockville, Md., June 1997, 20.
8. Gina Kolata, "Researchers Clone Embryos of Human in Fertility Effort," *New York Times*, 26 October 1993, A1.
9. T. Adler, "Bidding Bye-Bye to the Black Sheep?" *Science News*, 9 March 1997.
10. Martha Groves, "Transgenic Livestock May Become Biology's Cash Cow," *Los Angeles Times*, 1 May 1997 (Internet edition).
11. Gina Kolata, "Lab Yields Lamb with Human Gene," *New York Times*, 25 July 1997, A6.
12. Gina Kolata, "Lab Yields . . . ."
13. Ronald Hoffman, quoted by K. C. Cole, "Upsetting Our Sense of Self," *Los Angeles Times*, 28 April 1997.
14. George Johnson, "Don't Worry: They Still Can't Clone a Brain," *New York Times*, 2 March 1997.
15. C. Craig Venter, "Placing Cloning in the Context of the Genome," Conference on Mammalian Cloning: Implications for Science and Society, 26 June 1997, Crystal City Marriott, Crystal City, Virginia.

16. C. Craig Venter, "Placing Cloning . . . ."

17. James Thompson et al. *Genetics in Medicine* (Philadelphia: Saunders, 1986).

18. Incidence of these common genetic diseases is taken from the OMIM—Online Mendelian Inheritance in Man, web address: http://www3.ncbi.nlm.gov/omin/. See also Victor McKusick, *Mendelian Inheritance in Man: A Catalog of Human Genes and Disorders*, 11th ed., 2 vols. (Baltimore, Md.: Johns Hopkins University Press, 1994).

19. Michael Specter with Gina Kolata, "After Decades. . . ."

20. *The Sciences*, May/June 1997, 10.

21. Stephen Jay Gould, "Dolly's Fashion and Louis's Passion," *Natural History*, June 1997, 21.

22. David R. Collins, *Eng and Chang: The Original Siamese Twins* (New York: Dillon Press, 1994).

23. Geoffrey A. Machin, "Conjoined Twins: Implications for Blastogenesis," *Birth Defects: Original Articles Series* 20, no. 1, 1993, March of Dimes Foundation, 142.

24. C. R. Austin and R. V. Short, *Reproduction in Mammals*, 2nd ed. (New York: Cambridge University Press, 1993).

25. Elaine Landau, *Joined at Birth: The Lives of Conjoined Twins* (New York: Grolier Publishing, 1997).

26. I am indebted to Ona Marie Faye-Petersen of the UAB Department of Anatomical Pathology for my understanding of such matters.

27. Tim Friend, "Getting to the Nucleus of Cloning Concerns," Panel with Colin Stewart, Tom Murray, and Neil First, *USA Today*, 1 April 1997.

28. My discussion of the history of biology's understanding of human genetics is indebted to Kelly Cox Smith (Duke) doctoral dissertation, *The Emperor's New Genes: The Role of the Genome in Development and Evolution* (Ann Arbor, MI: UMI Dissertation Services, 1994). See also Glenn McGee, *The Perfect Baby: A Pragmatic Approach to Genetics* (Lanham, Md.: Rowman & Littlefield, 1997), 68–71.

29. Cited in Kelly Smith, *Emperor's New Genes* . . . 50. [Dalenpatius (de Plantage's pseudonym) (1699) "Extrait d'une lettre de M. Dalenpatius a l'auteur de ces nouvelles contenant une de'couverte curieuse, faite par le moyen du microscope" abstracted in Leeuwenhoek, *Philosophical Transactions of the Royal Society*, 21.]

30. "The Science and Application of Cloning," National Bioethics Advisory Commission, *Cloning Human Beings*, 32–33.

The False Seers of Assisted Human Reproduction

Human life is sacred. The good Lord ordained a time-honored method of creating human life, commensurate with substantial responsibility on the part of the parents, the responsibility to raise a child appropriately. Creating life in the laboratory is totally inappropriate and so far removed from the process of marriage and parenting that has been instituted upon this planet that we must rebel against the very concept of human cloning. It is simply wrong to experiment with the creation of human life in this way.

U.S. Congressman Vernon Ehlers, Michigan,
a former research physicist, urging Congress to pass a law
making human cloning a federal crime[1]

· · · · ·

This chapter discusses the recent history of erroneous pronouncements about the ethics of human reproduction. It is also a lesson about predictions of slippery slopes, especially from changes in human reproduction.

Beware False Prophets of Doom

Many important parts of the debate over human cloning are not new. Just twenty years ago, it is remarkable how many famous experts opposed help for the infertile and how, if it had been up to them, such help never would have developed.

Before and after the first birth in 1978 through in vitro fertilization (IVF), sensationalistic terms were often used. Eliminating a genetic disease such as cystic fibrosis was condemned as "genetic engineering." Creating a wanted baby by combining the husband's sperm and the wife's egg for a few hours outside the fallopian tubes was called "making test-tube babies" (what the father of the first such baby, John Brown, called "helping nature along a bit"[2]). Cloning human embryos was condemned by one of the pioneers of modern medical ethics, American theologian Paul Ramsey, as making "carbon copy people."[3] Doing anything different than accepting our genetic fate with our heads humbly down was condemned as "playing God," with connotations of hubris, arrogance, and blasphemy.

There has also certainly not been any direct relationship between the ability to achieve a scientific breakthrough and moral wisdom. Famous scientists frequently embarrass themselves in making predictions in bioethics. James Watson, who won the Nobel prize with Francis Crick for his work on DNA, predicted that dangerous events would follow Louise Brown's birth.[4] He feared that deformed babies would be born and that they would then have to be raised by the state in custodial homes, or that they might even be victims of infanticide.[5] Max Perutz, who was a colleague of Robert Edwards' at Cambridge—and, like Watson, a Nobel prize-winner—also condemned creating human babies by IVF:

I agree entirely with Dr. Watson that this is far too great a risk. Even if only a single abnormal baby is born and has to be kept alive as an invalid for the rest of its life, Dr. Edwards would have a terrible guilt upon his shoulders. The idea that this might happen on a larger scale—a new thalidomide catastrophe—is horrifying.[6]

Nor does working with pregnant mothers in medicine seem to bring the ability in reproductive ethics to see the forest for the trees. Obstetrics researcher John Marlow emphasized in 1978 that severely defective babies could be created, and that "the potential is there for serious anomalies should an unqualified scientist mishandle an embryo."[7] Another obstetrician said, "What if we got . . . a cyclops? Who is responsible? The parents? Is the government obligated to take care of it?"[8]

When will we ever learn whom we can trust and remember who has misled us before? Look back in 1977 at Jeremy Rifkin, who has made a living ever since by exaggeration and hyperbole and who then revved up the fear that Louise Brown might be psychologically "monstrous":

What are the psychological implications of growing up as a specimen, sheltered not by a warm womb but by steel and glass, belonging to no one but the lab technician who joined together sperm and egg? In a world already populated with people with identity crises, what's the personal identity of a test-tube baby?[9]

Jump to twenty years later. Predictably, Rifkin bolted out the gate in 1997 in the race to condemn human cloning, only now behind the cloak of his self-created "Foundation for Economic Trends." Rifkin demanded a world-wide ban on human cloning with penalties for transgressions as severe as those for rape and murder.[10] "It's a horrendous crime to make a Xerox of someone," he declared ominously. "You're putting a human into a genetic straitjacket. For the first time, we've taken the principles of industrial design—quality control, predictability—and applied them to a human being."[11]

Philosopher Daniel Callahan, a philosopher who had worked in the Catholic tradition and who founded the bioethics center known as the "Hastings Center," argued in 1978 that the first case of IVF was "probably unethical" because there was no possible guarantee that Louise Brown would be normal; he then said it would be ethical to proceed with IVF after this first healthy birth.[12] Callahan added that many medical breakthroughs are actually "unethical" because we cannot know (using the philosopher's strong sense of "know") that the first patient will not be harmed. Two decades later when asked about human cloning, it was no surprise that he opposed cloning, implying that it was unethical: "We live in a culture that likes science and technology very much. If someone wants something, and the rest of us can't prove they are going to do devastating harm, they are going to do it."[13]

Leon Kass, a social conservative who thirty years ago was trained as a biochemist and physician, and who teaches at the University of Chicago, argued strenuously in 1971 that babies created by artificial fertilization might be deformed: "It doesn't matter how many times the baby is tested while in the mother's womb," he averred, "they will never be certain the baby won't be born without defect."[14] Twenty-six years later, Kass predictably condemned cloning a human baby in the strongest possible terms, calling it a "horror" and "incest," "morally repugnant" and a "perversity." Writing after the announcement of Dolly's cloning, he claimed,

> Human cloning would represent a giant step toward turning begetting into making, procreation into manufacture (literally, something "handmade"), a process already begun with in vitro fertilization and genetic testing of embryos.[15]

Why should we trust the claims to wisdom of such people? Have they ever admitted they were wrong to oppose the creation of thousands upon thousands of babies created by IVF? That they misjudged the risk of harm to these babies? Are not the same premises still operative in their thinking that they assumed two decades ago, leading them now to similarly unsound conclusions?

Their arguments certainly do not seem very compelling. Although hindsight always gives 20/20 vision, what these critics overlooked was that no reasonable approach to life avoids all risks. Without some risk, there is no progress, no advance. Without risk, pioneers don't cross prairies, astronauts don't walk on the moon, and Freedom Riders don't take buses to integrate the South.

These critics of IVF demonstrated a psychologically normal but nevertheless illogical tendency to magnify the risk of a harmful but unlikely event. A highly unlikely result—even if that result is very bad—still represents a very small risk. An anencephalic baby (a baby born without any brains) is an extremely bad but unlikely result, but this small risk shouldn't deter people from having children.

Even if Louise Brown could not have been harmed by the external effects of being conceived in this way, by the reaction of others, or by being born with some medical defect, IVF could still have been wrong if an IVF baby could have been harmed simply by having been brought into existence through IVF. And some critics claimed just this.

Whether possible children can or cannot be harmed by in vitro conception is a philosophically interesting consideration, because it also applies to gestating a cloned human embryo. At the time of Louise Brown's birth, Religion professor Hans Tiefel argued, on the assumption that IVF subjected a baby to greater-than-normal risks, "No one has the moral right to endanger a child while there is yet the option of whether the child shall come into existence."[16] On this topic, consider the old Yiddish joke: 1st— "Life is so terrible! Better to have never existed." 2d—"True, but who is so lucky? Not one in a thousand."

In a perfectly understandable sense, a person cannot be harmed in being created by IVF because otherwise he would not have existed. The only way an IVF baby might plausibly be considered to have been harmed is if the IVF baby had some defect or deficiency that normal babies do not have. Because this had not been the case and no one had any reason to think so at the time, the objection is not a good one. (Philosophically, this question is more complicated than the above response allows. We shall consider it again in Chapter 9, where this kind of objection to human cloning is discussed.)

Critics of IVF also envisioned potential harms to society. *US News & World Report* described Louise Brown's birth as a "disturbing" and "ominous" event, an indication of how "science wields its growing power to decide who shall be born, how, and to whom."[17] An unstated assumption here was that science—not infertile couples—decides who is born. This is another instance of how many people (alas, usually men) make women invisible in these debates and keep forgetting that women must not only decide to gestate fetuses but also can legally decide to abort such fetuses if conditions warrant. No one but the female carrying the fetus can "decide who shall be born," because it takes her commitment to the use of her body for nine months to do so.

Many critics used the slippery slope metaphor, among them James Watson, who concluded that an international effort was needed to "de-emphasize" research that would circumvent "normal sexual reproduction." But we should always be suspicious about such predictions of vague harms in the future. Watson and other critics worried that IVF was too close to cloning humans, and seemed to think that the latter meant something like the scenario of *The Boys from Brazil.* Of course, the idea that humans who were originated by cloning would be automatons, mindlessly taking orders, was just as stupid then as it is now. If someone had the motive and if it were feasible, such a project would not require IVF or cloning. Some modern Mengele could kidnap a dozen street urchins from a ghetto of Rio de Janeiro, raise them in isolated areas, and achieve the same result—if it were possible to so easily create mindless humans of a certain sort. The problem is always that as adults they would not be obedient zombies but people with ordinary minds who could think for themselves.

Clone Furor: First Wave

In this period of the early 1970s, human cloning was also discussed and roundly condemned by the same people: Paul Ramsey, Leon Kass, and Jeremy Rifkin.

James Watson, in an article entitled "Moving Towards Clonal Man" in the *Atlantic* magazine in 1971, seemed unable to understand that the very negative reaction he was predicting in others about Louise Brown's birth was being caused by his own writings: "we must anticipate strong, if not hysterical, reactions from many quarters," he wrote in strong, somewhat hysterical terms.[18] Implying that human cloning was a very real possibility in 1971, Watson thought that for modern humans, creating a person by cloning would be "a potential threat to their basic way of life" and added despairingly that, upon hearing of a person originated this way, "the first reaction of most people to the arrival of these asexually produced children, I suspect, would be despair." He thought human cloning "most likely will appear on the earth within the next twenty to fifty years," and need not involve a totalitarian state because (in a stunningly sexist remark):

the boring meaninglessness of the lives of many women would be sufficient cause for their willingness to participate in such experimentation [gestating a fetus originated by cloning], be it legal or illegal.[19]

Over and over again, Watson was guilty of making self-fulfilling doomsday prophecies. He predicted that if human cloning occurred, "The nature of the bond between parents and their children, not to mention everyone's values about the individual's uniqueness, could be changed beyond recognition. . . ." Worse (for him!), "experiments on cell fusion might no longer be supported by federal funds or tax-exempt organizations." So he wanted to prohibit any attempt at human cloning and to make illegal any research on "experimental work with human embryos." Anticipating the current objection that if the United States and Britain ban human cloning, some other nation might permit it, he called for an international effort to ban both human cloning and research on cloned human embryos.

In all this, Watson made numerous assumptions that were entirely unfactual and which, even at the time, had lots of facts marshalled against them. For example, he assumed that infertile couples were a "potentially militant lobby" in politics, when in fact they have been almost impotent and have always been completely disorganized. He assumed as axiomatic that a child born with great artistic talent like Picasso would wish to have never been born because of unrealistic expectations placed on him to produce great art. Even if a plausible fear, such a claim must not just be assumed but empirically tested.

In a similar vein in 1972, Marc Lappe, an experimental pathologist, wrote "Risk-taking for the Unborn: The Ethics of In Vitro Fertilization."[20] Among other exaggerations, he compared creating babies through IVF to the risks to girls of mothers who took DES (diethylstilbestrol).

Both Watson and Lappe seemed to often impute bad motives to parents. More paradoxically, they also seemed to impute such motives to scientists like themselves. This is like the physician who urges us to trust him but who vigorously claims that other physicians can't be trusted.

Clone Furor: Second Wave

In 1978, David M. Rorvik wrote a sensationalistic book, *In His Image: The Cloning of a Man.*[21] In its opening, Rorvik revealed in hush-hush, breathless prose that an elderly, unmarried millionaire industrialist named Max had telephoned him at "my cabin on Flathead Lake in the mountains of western Montana" and convinced Rorvik that Max was going to clone a human baby of himself from a single cell of Max's adult body. Rorvik said that Max allowed him to meet "Darwin," the infertility specialist involved, to see the lab where Darwin and his assistant worked on the problems of nuclear transfer and matching the correct stage of cell division, to meet "Sparrow," the host mother, to watch as amniocentesis was done on Sparrow, and (in the final page) to testify that a baby had been born and that "blood analyses and other tests involving histocompatibility factors have demonstrated to Max's complete satisfaction that the child is indeed his clonal offspring."[22]

Rorvik was a science writer, but what did he really see? What did his seeing really prove? How could Rorvik ever know if such a boy was originated by cloning or merely Max's child of "Sparrow" by insemination? Max's own child produced this way could easily have the same blood type as his own.

In 1982 in a Federal District Court in Philadelphia, a judge ruled that the book was "a fraud and hoax," quite an embarrassment to Lippincott Publishing in Philadelphia, which is ordinarily a respectable publisher of medical texts and books, and which had issued a special endorsement of the book in its preface.[23] The hoax set off a furor of ethical condemnations. As Ruth Macklin notes, the book "triggered the first wave of 'clone furor,' which included articles and ethical analyses by philosophers, scientists and others."[24]

Because of this fraud and the one previously described of the cloned mice, as well as some other hoaxes claiming to have produced an IVF baby before 1978, it is easy to understand why, when Louise Brown was born in 1978, many people did not believe she had been created by IVF. Similarly, Ian Wilmut has his skeptics.

Clone Furor: Third Wave

In 1993, fifteen years after the birth of Louise Brown, another brouhaha occurred when some well-known American bioethicists were hyperventilating for a few days when the *New York Times* reported, on the first page of its widely read Sunday edition, that scientist Jerry L. Hall had "cloned" 17 human embryos into 48 as a way of increasing the supply of embryos in fertility clinics.[25] The next day, other newspapers, CNN, and *Good Morning America* took the *Times* story at face value, as did several bioethicists. Jeremy

Rifkin instantly condemned such cloning, claiming that "this is the dawn of the eugenics era."[26] Daniel Callahan intoned that cloning could violate a right "to our own individual genetic identity," which he claimed that each human had.[27]

To its credit, *Newsweek*, which had made a few mistakes in its 1978 stories about Louise Brown, got the story right in 1993. Here is what *Newsweek* said:

> It is one of the most sought-after coups of 20th century journalism, along with the identity of Deep Throat and Senator Packwood's diaries—the first story that can plausibly use "human" and "clone" in the same headline. . . . Last week, *The New York Times*, [on the basis of] an apparent misunderstanding of a paper reporting a technical advance in embryology, touched off an echo of the same hysteria with a page-one story whose headline suggested that human embryos were being cloned in a laboratory. Within days medical ethicists were gravely measuring the slipperiness of the slope on which humanity now teetered, while demonstrators marched outside laboratories insisting that no one would ever clone their DNA.[28]

What most people understood by cloning was what Woody Allen had imagined in his movie *Sleeper*, reproduction of an identical physical copy by nuclear somatic transfer from a cell of a human body (in *Sleeper*, the cells from a dictator's nose); or what was depicted in the more recent movie *Jurassic Park*—recreation of extinct species from preserved DNA (not possible at present).

The *New York Times'* article had described copying of undifferentiated cells of embryos by twinning, a technique that is common in animal husbandry and was easily applied to human embryos. At the time in 1993, what this reporter and various bioethicists did not understand was that once human cells specialized into skin cells, blood cells, and so on, they could not be returned to their undifferentiated state to create a copy of the entire body of their donor. From the cells of the dictator's nose, you get . . . a nose! After the *Newsweek* article, the story simply disappeared. *Newsweek* noted that the *Times* "did not respond to numerous requests for comment last week."

This "fact" of course is exactly what Willadsen and Wilmut later proved to be wrong. Willadsen and Wilmut revealed in 1997 that they had done what everybody incorrectly thought had been done in 1993.

Into this non-issue in 1993 jumped many unwary bioethicists. Good sound-bites and memorable phrases get a bioethicist quoted, whether it's a local paper or a national one. If you want to make a big splash, you have to say that human cloning is as dangerous as North Korea launching a nuclear attack on Canada.

Bioethicist Art Caplan decried the Dorian Gray scenario, where in Oscar Wilde's play, Dorian never ages, but can periodically see what he should look like at his age by looking in a mirror.[29] Similarly, he claimed the cloned child could be harmed by seeing, at 40, what he will look like at 60. Second, "When you deliberately set out to make copies of something, you lessen its worth," he said. He also claimed that "creating embryos solely for the purpose of genetic diagnosis is morally suspect."[30] Caplan, who is perhaps the quickest bioethicst in American to comment on a breaking ethical issue, repeated his "Dorian Gray" objection to human cloning in 1997.[31]

Bioethicist Ruth Macklin, who herself briefly succumbed to hyping clone fears and who quickly recovered from this occupational disease, reviewed such comments by bioethicists in a scholarly article in the fall of 1994.[32]

What is, perhaps, surprising about the bioethicists' responses [to the twinning story] was the tendency to envisage or even contemplate worst-case scenarios. Margaret Somerville, from McGill University, was quoted as saying: ". . . what we are talking about is the ability to mass-produce humans."[33] The article in *Time* noted that, "Ethicists called up nightmare visions of baby farming, of clones cannibalized for spare parts."[34] Boyce Rensberger in the *Washington Post* blamed ethicists for parlaying figments of their imagination into hypothetical scenarios, "bizarre scenarios of raising armies of clones or of creating twin siblings to harvest for organs."[35]

In 1997, testifying before NBAC, Ruth Macklin and law professor John Robertson were the only bioethicists who showed any common sense in urging caution before enacting a ban on cloning, a sentiment that has been started by NIH Director Harold Varmus.

Unfortunately, physicians who were close to the needs of infertile couples in 1994 were no less dramatic. Mark Hornstein, director of the Brigham and Women's Hospital in Boston, implied that cloning human embryos was like Nazi-Mengele eugenics.[36] Dr. Mark Sauer, medical director of the IVF center at Chicago's Michael Reese Hospital said, "It's a dangerous turn, trying to create the perfect child and then duplicating it. What do you do if you don't like the first child? Throw the cloned embryo away?"[37] (But four years later, Sauer had a change of heart and pleaded against a ban on use of cloned human embryos in assisting infertile couples.[38])

In fairness, some bioethicists pooh-poohed these Jeremiahs and tried to calm the media's fires. University of Wisconsin bioethicist and pediatrics professor Norman Fost led this response, and Albert Jonsen, Ruth Chadwick, and Ruth Macklin soon joined in. But at the time, their voices were lost in the wilderness, and when it was realized that no one had cloned humans in the important sense (but in the trivial sense of twinning), few people remembered which bioethicists had been so incorrect so publicly.

In April of 1997, a 63-year-old woman was revealed to have gestated a human baby created from her husband's sperm and the donated egg of a young woman. All over again, we had to hear the cries of the sky falling because an older woman had helped create a human child. Meanwhile, actor Tony Randall at 77 was going on talk shows with his young wife and new baby, with no one throwing stones at him.

The National Bioethics Advisory Commission (NBAC)

Given this history, it is interesting to comment on the National Bioethics Advisory Commission (NBAC) and whom it chose to invite to address its deliberations. I think it important to stress that this was not a commission appointed specifically to study the ethics of human cloning (but to study the ethics of the new genetics, among other topics), because if NIH had wanted to do that, it would have gotten experts in reproductive ethics, embryology, reproductive law, and bioethicists known for their wisdom about reproduction, making sure that both sides of the ethical coin were adequately represented. Instead, this was a commission that already existed and that President Clinton used to create a quick report for Congress and his own legislative agenda.

NBAC was given 90 days in late February to study human cloning and to deliver recommendations. Its members never had a doubt that they were going to oppose human cloning. It is obvious why from their professions and orientations.

Of the 17 members, there was not one strong voice known publicly for his or her defense of human reproductive rights or known for his or her skepticism about previous tempests-in-teapots in bioethics. There were 7 women and 10 men. There were 5 M.D.'s, two of whom (Bernard Lo and Ezekiel Emanuel) had expertise in clinical medical ethics, i.e., expertise very much focused on particular contexts of actual medical care, but not so much on more general philosophical analysis. There were two law professors (Alex Capron and Alta Charo) plus one M.D. with a law degree (the director of the Hawaii State Department of Health). There was also one professor of: psychology, nursing, and religious studies (Jim Childress). There was an advocate for the mentally ill, a consumer president of a Virginia bioethics network, a business officer of a pharmaceutical company, and a scientist from Cold Springs Harbor. There were two pediatricians. There were two official bioethicists, Patricia Backlar from Portland State University (Philosophy/Medicine) and Tom Murray, Case Western Reserve University. There was a famous senior physician, Eric Cassell, and an M.D./Ph.D. (Genetics) from Stanford Medical School.

This Commission needed to have been "Fletcherized" by having someone like the late Joseph Fletcher or the existing John Fletcher (no relation) at the University of Virginia voice the pro-cloning side. Where is our American Peter Singer when we need him?

For reasons unknown, the Commission invited many people from religion to testify. When they invited Leon Kass, they heard a condemnation that was perfectly predictable but without mention of his previous predictions that have proved false. Gilbert Meilaender, who had opposed withdrawing food and water from dying patients as immoral in the 1980s on symbolic religious grounds, predictably opposed all human cloning on the same grounds.

This bias was reflected in the huge section of the NBAC *Report* on "religious perspectives." What is odd about so much emphasis on religious views, especially since most of these views were against human cloning for the usual vague reasons, is that scientific policy is usually not based on such considerations. The only exception is matters that concern abortion (this topic may have been linked in the minds of some Commissioners to human cloning, especially because two of the Commissioners had also served on the 1994 Human Embryo Panel).

Nevertheless, the whole emphasis on religious views was quite odd. It would be as if, when President Clinton proposed a national medical insurance in 1994, there had been a national commission and its report had placed a great emphasis on religious views. Strikingly, such views were absent in many of the discussions of physician-assisted dying that were occurring during the cloning controversy.

Conclusion

Over the last twenty years, we have seen numerous changes in human reproduction. Each has been greeted with alarm and fear. Each has been predicted to ruin human civilization as we know it. Each has had remarkably little effect on anyone other than the immediate family of those involved, none of whom have ever claimed in retrospect that they were harmed.

Physicians, distinguished scientists, and some well-known bioethicists have done very poorly in helping reduce the public's fears. A few self-appointed public intellectuals have usually done just the opposite, fanning the flames of public fear and promising terrible slippery slopes. Physicians have done little better when they talk to reporters about breaking news in reproductive ethics.

Indeed, there has been an unfortunate tendency recently for a few very public bioethicists to preach to the public as if looking down from Mount Sinai on all breaking and not-so-breaking events. Even before Dolly's announcement, several prominent bioethicists had already become increasingly strident and absolutist that laws permitting physician-assisted dying would be America's ethical Armageddon. When the news of the cloning of Dolly broke, each swung out his yardstick, gunslinger-style, to measure how far down the slippery slope we had tumbled.

It is sad to have to review this history, but it is necessary to do so. This story would be funny if it didn't block important genetic research or prevent promising work with human embryos. But as we shall discuss later in Chapter 7, the prospects are quite dismal for even lifting the continuing ban on medical research on embryos, and if that is so, imagine how far we would have to go to allow an infertile couple to originate a child by cloning.

Notes

1. *Congressional Record-House*, 4 March 1997, H713.
2. John Brown and Lesley Brown, *Our Miracle Called Louise* (London: Paddington, 1979), 98.
3. Paul Ramsey, *Fabricated Man: The Ethics of Genetic Control* (New Haven, Conn.: Yale University Press, 1970).
4. *Newsweek*, 7 August 1978, 69.
5. James Watson, "Moving towards Clonal Man," *Atlantic*, May 1971, 53.
6. Patrick Steptoe and Robert Edwards, *A Matter of Life: The Story of a Medical Breakthrough* (London: Morrow, 1980), 117.
7. *U.S. News and World Report*, 7 August 1978, 24.
8. *Time*, 31 July 1978, 59.
9. J. Rifkin and T. Howard, *Who Shall Play God?* (New York: Dell, 1977), 15.
10. Ehsan Masood, "Cloning Technique 'Reveals Legal Loophole'," *Nature* 38 (27 February 1987).
11. Jeffrey Kluger, "Will We Follow the Sheep?" *Time*, 10 March 1997, 70.

12. *New York Times*, 27 July 1978, A16.

13. Knight-Ridder newspapers, 10 March 1997.

14. Leon Kass, "The New Biology: What Price Relieving Man's Estate?" *Journal of the American Medical Association* 174 (19 November 1971), 779–788.

15. Leon Kass, "The Wisdom of Repugnance," *The New Republic*, 2 June 1997.

16. Hans Tiefel, "In Vitro Fertilization: A Conservative View," *Journal of the American Medical Association* 247, no. 23 (18 June 1982), 3235–3242.

17. *U.S. News & World Report*, 7 August 1978, 71.

18. James Watson, "Moving . . . ," 51.

19. James Watson, "Moving . . . ," 52.

20. Marc Lappe, "Risk-taking for the Unborn: The Ethics of In Vitro Fertilization," *Hastings Center Report* 2, no. 1 (February 1972), 1–3.

21. David M. Rorvik, *In His Image: The Cloning of a Man* (Philadelphia, Pa.: J. B. Lippincott, 1978).

22. David M. Rorvik, *In His Image*, 207.

23. Peter Steinfels, "Latest Advances in Cloning Challenge Bioethicists," *New York Times*, 30 October 1993, A7.

24. Ruth Macklin, "Splitting Embryos on the Slippery Slope," *Kennedy Institute of Ethics Journal* 4, no. 3 (September 1994), 217.

25. Gina Kolata, "Researchers Clone Embryos of Human in Fertility Effort, *New York Times*, 26 October 1993, A1.

26. Philip Elmer-DeWitt, "Cloning: Where Do We Draw the Line?" *Time*, 8 November 1993, 65.

27. Philip Elmer-DeWitt, "Cloning . . . ," 65.

28. Jerry Adler, Mary Hager, and Karen Springer, "Clone Hype," *Newsweek*, 8 November 1993, 61.

29. Gina Kolata, "Cloning Human Embryos: Debate Erupts Over Ethics," *New York Times*, 27 October 1993, A1.

30. Rebecca Kohlberg, "Human Embryo Cloning Reported," *Science* 262 (1993), 653.

31. Art Caplan, "Ethical Issues," Conference on Mammalian Cloning: Implications for Science and Society, 26 June 1997, Crystal City Marriott, Crystal City, Virginia.

32. Ruth Macklin, "Splitting . . . ," 214.

33. Rebecca Kohlberg, "Human Embryo . . . ," 653.

34. Philip Elmer-DeWitt, "Cloning . . . ," 65.

35. Boyce Rensberger, "The Frightful Invasion of the Body Snatchers Will Have to Wait," *Washington Post* , 1 November 1993, A3.

36. Gina Kolata, "Cloning Human Embryos . . . ," A1.

37. Gina Kolata, "Cloning Human Embryos . . . ," A1.

38. Gina Kolata, "For Some Infertility Experts, Human Cloning Is a Dream," *New York Times*, 7 June 1997 (national Internet edition).

Misconceptions

We simply cannot stand by and allow humans to be copied. That would be breaking through an ethical barrier that goes far beyond even the barrier of the atomic bomb.

Juergen Ruettgers, German research minister[1]

The present and pressing question before us is human cloning—the seriousness of which is obvious to all. This is not research to improve "salad oil," nor research to improve "motor oil," this is something new and gravely important because the focus is human—the object, the subject, the means and the stakes are human!

. . . In the process of "making" a clone, just who is in charge of quality control? What qualifications are relevant to quality controllers . . . ?

Cloning is not now and never will be a poor people's campaign. Could it be or become an entitlement requiring public subsidy? This is a most undesirable shift because it replaces ethical categories with manufacturing imperatives.

John Cardinal O'Connor, Head of the New York Archdiocese of the Roman Catholic Church[2]

· · · · ·

Eighteenth-century philosopher and physician John Locke, in the "Epistle to the Reader" in his *Essay Concerning Human Understanding*, writes that when it comes to great ideas, "The commonwealth of learning is not at this time without master-builders" such as the likes of his contemporaries, Robert Boyle, Thomas Sydenham, and Isaac Newton. But for Locke himself, it was "ambition enough to be employed as an under-laborer in clearing the ground a little and removing some of the rubbish that lies in the way of knowledge."[3] In thinking about cloning humans, it is also necessary to clear the ground a little.

Our Legacy from Science Fiction

Many misconceptions about human cloning come from science fiction. At least half of this fiction sees human clones as little better than troglodytes.

The six cloned sisters on *Battlestar Galactica* create an indelible impression, as do sinister stories about cloning on *The X-Files*. It is hard to get past these impressions, or the toad-human clones on *Lois and Clark*. People who create humans by cloning in fiction have evil motives and their creations cause evil. Take three influential examples.

Bladerunner, the 1982 science fiction classic by Ridley Scott, is a paradigm (ironically in relation to Dolly, this movie was based on a novel by Philip K. Dick entitled, "Do Androids Dream of Electric Sheep?"). In *Bladerunner,* the Tyrell Corporation by 2019 has so successfully cloned humans ("replicants") that they can only be distinguished from humans by elaborate tests for emotional responses by checking eye movement, breathing, and pulse rate while they respond to sensitive questions. Although designed without human emotions, replicants have evolved such emotions. Replicants were designed to be "superior in strength and ability, and at least equal in intelligence," not to normal humans, but to the brilliant genetic engineers who created them.

The moral message comes fast: replicants are slave labor in the hazardous conditions of off-world planets, exploited for exploration there. On meritocratic or egalitarian principles, clones should possess equal rights with humans because they are superior in strength, ability and intelligence. But replicants are treated worse than pets, more like things. Understandably, replicants revolt and humans come to fear them. For replicants to return to Earth is illegal, on pain of death. Without hearing or trial, any replicant on earth can be killed by Bladerunners, a special police unit. When replicants are so executed by Bladerunners, they are said to be "retired."

The society that has done all this is depicted as, quite simply, hell. Los Angeles in 2019 is an ecological nightmare. The city appears chaotic, inhumane, dominated by a few large corporations, and with a language of the streets called Cityspeak, a pidgin polyglot. Has cloning humans created all this? Or has the society that allowed human cloning created all this? Either way, this apocalyptic Los Angeles is something to avoid.

Cyteen, C. J. Cherryh's classic 1988 trilogy about human cloning, has similar themes. Probes transport humans to distant stars, but it can take half a lifetime to get there and back, so most stay. Over three centuries, the off-world peoples grow culturally distant from the people of Earth, who are beset by overpopulation, ancient ethnic disputes, and parochial concerns. So the habitable world Cyteen, founded by dissident engineers and biologists, declares its independence from Earth.

There is a need in this cradle of new civilization for many more humans than could be transported by spaceship, so the Reseune Corporation develops artificial womb tanks to grow cloned humans called "Azis." When Earth declares war,

> Reseune bred soldiers, then, grim and single-minded and intelligent, oh, yes, bred and refined and honed, knowing by touch and reflex what they had never seen in their lives, knowing above all what their purpose was. Living weapons, thinking and calculating down one track.[4]

Of course, the Azis develop in unexpected ways. Some citizen-humans have Azi-twins and love them as human twins. Predictably, such Azis are treated by some humans as things or slaves, and the moral tone of *Cyteen* emphasizes the injustice of such treatment.

The lasting impression from these two famous, fictional treatments is that humans originated by cloning will be exploited, treated as serfs or slaves, and never be the equal of real humans. Given this influence, as well as that of dozens of other science fiction novels/movies that feature cloned humans, it was understandable that, a month after the announcement of his cloning of Dolly, Ian Wilmut would say that human cloning would "violate a right because the clone would not be treated as an individual."[5] In all our science fiction stories, the bad guys always treat human clones unjustly.

Kate Wilhelm's great 1974 novel, *Where Late the Sweet Birds Sang*, written at the height of the first wave of clone furor, explores the tension when one human attempts to forge an individual "self" separate from her identical half-dozen twins.[6] Genetically-identical, gestated identically in womb-tanks, and raised under identical conditions, these siblings cannot stand to be apart from each other and feel traumatized when any one of their group-identity is missing. Wilhelm's novel assumes that it makes sense to talk of multiple copies of a genotype who are exactly alike and who must therefore think as a group, that human clones would be poor at thinking for themselves, and that particular types of humans could be created for slave-like jobs.

Similarly, in Fay Weldon's *The Cloning of Joanna May*, nefarious, sexist motives are at work when a rich man wants to create females like his young wife, Joanna.[7] He creates four copies of Joanna's genotype by using parthenogenesis on four of her eggs (stimulating an egg to divide and start differentiation, an unlikely event) and then implants the resulting embryos in women to be gestated.

David Brin, a scientist who writes science fiction, paints the most positive picture of multiply-cloned humans in *Glory Season*.[8] This novel of fantasy paints a picture of a matriarchy where men are peripheral sailors and only allowed to mate during one season in order to keep the gene pool healthy. Most women are cloned from a particular type, which over time has become an economic guild known for its unique talents, such as working with numbers or metal-working.

Perhaps the most famous movie about human cloning was based on Ira Levin's *The Boys from Brazil.* The 1978 movie of the same name has a Jewish Nazi-hunter named Lieberman (modeled on Simon Wiesenthal) who stumbles upon a plot in Paraguay that is the brainstorm of infamous Nazi physician-experimenter Josef Mengele (who was still alive there when the movie came out). Lieberman, played by Laurence Olivier, discovers that 94 embryos were created by cloning cells containing the genes of Adolf Hitler. A neo-Nazi group had 94 women gestate these embryos to birth, whereupon they were adopted by couples who matched the ages and personalities of Hitler's parents. When the little Adolfs are teenagers, lethal accidents to their fathers mimic the death of young Adolf Hitler's father. Upon finally meeting the boy, Mengele assumes the boy will think and feel like the real Adolf Hitler, but instead the boy is attached to his adopted father, whom Mengele has just killed. Nevertheless, the boy shows a vicious streak by allowing his Rottweilers to rip Mengele apart for killing his father.

In this movie, the key biological event is fudged. How the embryos got to be babies and then adopted is not really described (there is a kind of brief, dreamy reminiscence by Mengele). The movie ignores the role and choice of real women in human procreation.

The movie also assumes a simplistic genetic determinism. All the boys are identical in personality, despite the fact that even the best attempts to match 94 boys to 94 adoptive couples would produce great variations in parenting, local friends, educational systems, extended families, national loyalties, and exposure to media.

The Human Research Embryo Panel in 1994 (see Chapter 7) correctly concluded that, "Popular views of human cloning derive from science fiction books and films that have more to do with cultural fantasies than actual scientific experiments."[9] However, George Annas in his Senate testimony faulted this panel and another for ignoring the lessons of science fiction. "Both of these expert panels were attempting to sever science from its cultural context. Literary treatments of [human] cloning help inform us that applying this technology to humans is too dangerous to human life and values."[10]

Science fiction is no better in most other movies and novels that mention human cloning. From all of these movies, one can infer that: (1) at first only dictators will be cloned (Woody Allen's *Sleeper*, Hitler in *Boys*); (2) Human clones will gestate in womb tanks until they mature (Kate Wilhelm's *Where Late the Sweet Birds Sang*); (3) Clones will grow from zygotes to adult size in a few days, or sometimes, instantaneously (*Multiplicity*); (4) On emerging from their cylinders, or pods, the fully grown clones will be emotionless, murderous zombies, and act like they've had lobotomies (*Invasion of the Body Snatchers*); (5) People originated by cloning will always be originated as multiples raised in the same environment at the same time, never solely as individuals or as individuals in different times and places (Pamela Sargent's *Cloned Lives*, plus all other books and novels on this list); (6) Cloned women will be tall, thin, and beautiful (*Stepford Wives*); (7) If a good man or woman happens to be cloned, the clone will be evil (*Dr. Jekyll and Mr. Hyde*); or the good man will go to prison for crimes committed by the evil clone. Or the good twin will take up the life left behind by the evil twin (Bette Davis, *A Stolen Life*); (8) The scientist who clones a mammal will be killed by a clone, usually in his own lab-environment (*Jurassic Park*).

Our mythology has been also strongly biased against twins and triplets. Very few twins are as famous as Abigail van Buren and Ann Landers (although isn't it interesting that both ended up in the same niche in the same profession?). Who are the famous people in pre-modern times who were twins? It is hard to name any, probably because a twin was often killed. (Figures from myth don't count.)

There is a strong theme in history and mythology of the evil twin.[11] Probably this dates back to concerns about entailment, where the firstborn twin inherited all the land and his twin, born several minutes later, got nothing. The second twin would be resentful and would be feared as plotting to kill the firstborn. Concerns about twins may also be related to scarcity of resources, whether of food or land. In the biblical story of Isaac's fraternal Esau and Jacob, Jacob (the second born) and his mother conspire to deceive Isaac for the inheritance. Other twins in the Bible were Pharez and Zarah, Jacob's grandsons, who also fought over birthright. Among the Iroquois, one twin was seen as good, the other as evil, so the latter was killed. The most famous mythological story of feuding twins is that of Romulus and Remus, where in a power struggle, Romulus kills Remus.

Twins are also linked in literature to the theme of "the double" or "doubling," where the double of the normal person is allowed to express the normal's dark side. So the extra, second twin is the natural bad side to be feared. It is easy to see how twins and multiples of the same genome come to be associated in science fiction and literature with things to be feared.

Making Women Visible

Somewhere along the line, cloning got mixed up with artificial wombs, and these wombs in turn got mixed up with "genetic engineering." (It goes without repeating that persons originated by cloning could not be grown in test-tubes, but writers occasionally still use these terms.) But all these things are separate and separable. Although it is easy to leap from one topic to another as ideas, the leap in reality usually takes a long time.

We might call this the ectogenesis misconception. Not only can you not grow humans in vats, you can't grow them in artificial wombs because such wombs do not exist and will likely not exist anytime in the near future. That's a shame, because if anything like such a womb were possible, we could save many premature babies who now die. At present, if a "premie" does not have the ability to breathe (which fetuses get around 22–24 weeks), there is very little chance that a neonatal intensive care unit can save it. Billions of dollars and millions of doctor-hours have gone into trying to create such an artificial womb, but it's not here now and it's probably a long time off.

Furthermore, and for all kinds of reasons, originating a person by cloning is not going to be a common procedure among human couples. What this means is that women are not interchangeable machines in gestating human embryos and fetuses. A cloning factory for human beings is pure fantasy.

Recall that in *The Boys from Brazil,* how the 94 embryos from Hitler's cells were gestated was left vague. That is because this crucial detail would make the movie much less realistic if some attempt were made to explain it. How was this presumably done? Did Mengele and his men in Paraguay abduct 94 young women from some Indian village? Surely their mates and fathers would put up a fight and come after them. Were they made into prisoners for nine months and inseminated under anesthesia? What's to prevent such angry prisoners from trying to abort their fetuses, e.g., by suddenly jumping off beds or walls?

At the cloning conference, one speaker (from the animal/drug industry) objected to human cloning because it would require hundreds of women acting as surrogate mothers and "we can't exploit women like that." The presumption behind the objection, that hundreds of women would agree to gestate a genetically-identical embryo in some kind of gestational "factory" was so ludicrous that the concern about "exploitation" lost its force. We are not going to treat human females like hens or sows because human females have rights as persons, including rights to say how and whether they want to reproduce.

In real life, you would have to pay or persuade hundreds of women to gestate such embryos to get 94 certain births. And you would have to make sure that none of them wanted to keep their babies. But the main thing is that in *The Boys from Brazil* women are incidental incubators, not real persons who make choices about having babies. This lack of attention to the role of women has created a common misperception that humans originated by cloning could somehow be created without the active, ongoing choices and cooperation of women.

Some people have the bizarre idea that no one would bear any responsibility for raising a person originated by cloning other than the ancestor of the genotype. Untrue! For better or worse, the woman gestating the child will bear great responsibility, and if a male-genotype donor changes his mind and walks away from the pregnancy, the woman will be stuck with the baby. She will always be free to abort up until viability, just as if she had gotten pregnant by having had sex with the same man.

It is perhaps understandable that some women fear that any talk of human cloning is sexist because so many people are so flippant about the importance of their role in gestating a human baby. One thing we can know is that most of the people talking so flippantly have not actually carried a baby.

In modern democratic societies with a tradition of individual liberties, pregnancy for any reason is simply not imposable on women by the State. Given controversies about cloning, a fortiori, it is not going to be imposed on women through cloning. Human pregnancy is voluntary, and ought to remain as voluntary as possible.

They Would Be People

A clone is not a drone. Cloned humans would be people. It is a widely-accepted, general principle of modern philosophical ethics that people should be treated equally as moral agents unless there is a morally relevant reason to treat them otherwise. Every person should be treated with respect and as possessing equal moral worth until it is proven that he or she deserves to be treated otherwise.

This principle stems from the acceptance of Kantian, utilitarian, and Judeo-Christian theories of ethics that hold that impartiality is central to ethics. As some might say, from the point of view of the universe, no person's suffering should count any less than any other's.

From a practical point of view, acceptance of this principle means that treating people unequally requires passing a test of justification. The onus of proof is on anyone who would treat a cloned human unequally. Anyone who would treat persons of group A unequally from those of group B must specify some morally relevant difference between the two groups. Historically, we have had to learn quite painfully that skin color, religious belief, ethnicity, gender, and sexual orientation are not such differences.

This very strong moral principle entails a sub-principle that society should not discriminate against people according to their origins. "People are people," and it should not matter how they came to exist. Call this the "Principle of Non-Discrimination by Origins."

This is a surprisingly difficult idea for some to understand:

"[Human cloning] would be perhaps the worst thing we have ever thought of in the maltreatment of our species. It would be a kind of new slave class. You would have human beings who were made by other human beings for their purposes." (Nigel Cameron, theologian, bioethicist, and provost, Trinity International University, Deerfield, Ill.[12])

"It is not at all clear to what extent a clone will truly be a moral agent." (Leon Kass, bioethicist and professor, the University of Chicago).[13]

If you have a prejudiced reaction to a person, you cannot cite your own prejudice as a moral justification of why that person will be treated badly. This is like a person saying that there should not be racial integration because "those other people" will never accept it.

This principle of non-discrimination by origins means that no one should suffer any prejudice because of how he was created. Whether a child originated because of unmarried parents, one parent and an unwanted pregnancy, in vitro fertilization, gametic intrafallopian transfer (GIFT), as a twin, triplet, or quadruplet, or quintuplet should not matter. If a child was created as a result of multiple embryo implantation during assisted reproduction, or by cloning, how a child gets into this world does not make him less a person. Instead, we should judge a child by the same criteria that we use to judge any other person.

It took a long time in human history to accept this principle. For millennia, the cultures of Western civilization would not accept children of unmarried women as beings with normal rights: they could not enter synagogues, marry, inherit property, and sometimes, vote. To say that bastards were socially stigmatized is to use a euphemism. Unless you were a king, your illegitimate children were not invited to the important family gatherings at birth, marriage, death, and holiday meals.

Today, we realize that children who were not born to two, married, heterosexual parents had no control over their origins. Once they arrive into the world, such children must be accepted as persons with all the normal rights.

Legally, humans become persons at birth. The U.S. Constitution has been consistently interpreted in the last fifty years to hold that at birth a human can inherit money, be a tax deduction, and, if killed, be the subject of a homicide charge. Persons originated by cloning would not be slaves or sub-citizens. As law professor Lori Andrews notes, the 13th Amendment would protect them: "Neither slavery nor involuntary servitude, except as a punishment for crime whereof the party shall have been duly convicted, shall exist within the United States, or any place subject to their jurisdiction."[14]

At one time, it was feared that society might treat children born of in vitro fertilization as second-class citizens. Fortunately, children created this way have not been subjected to any special discrimination or ill-regard. Indeed, because they were so desperately wanted, they may be more loved than ordinary children, and the idea that such children would be looked upon by their parents as inferior is laughable. Nevertheless, it took a few minutes to educate the curious: when Lesley Brown took her baby Louise outside after her famous birth, she later said that wary neighbors peered into her baby carriage, expecting to see something "red and wrinkly," like the little monsters in the movie *Gremlins*.[15]

Obviously, many popular ideas about cloned humans are not in the world of real ideas. We have already indirectly disposed of the idea that persons originated by cloning would be slaves, factory drones, automatons, sub-human, or necessarily second-class citizens. Instead, they would just as much be persons as children born from in vitro fertilization. They would be gestated by normal women over nine months. They would be raised by normal parents in normal neighborhoods. The only difference between them and other children is that they would inherit one set of (chosen) genes rather than a randomly mixed set.

There is another silly idea that must be dismissed. Because humans originated by cloning are persons, it follows that we cannot kill such persons for their organs. This would be no more ethical or legal than knocking out your brother, transporting him to a hospital, anesthetizing him, and taking out one of his organs for transplantation. Simply because a person is originated in a new way does not mean that, to use Kant's classic terms, he can be used as a "means" to the good of others. Instead, he will be an end-in-himself with the same rights as any other person.

A variation on the theme of spare parts recognizes what bioethicists call the cognitive criterion of personhood. In this imaginative version, the body originated by cloning is never allowed to develop a sense of self or consciousness and is kept alive to be a perfect match for a person whose organs will probably fail one day. Since the cloned humanoid is not a person in the first place, this argument goes, wouldn't it be permissible to keep him alive as a source of future organs for the genetic ancestor or for others?

Here the question has been begged that you could do something to human bodies originated by asexual reproduction that you couldn't do to bodies originated by sexual reproduction. And you certainly cannot stop the development of a normal human baby after birth by giving it a lobotomy and then allowing the brain-injured body to grow to adulthood for organ parts. That is called murder.

To squash this idea from another direction, consider how much outcry there is when a surgeon crosses an ethical line in trying to create more transplantable organs for dying patients on waiting lists. When terminal patients are not resuscitated after a heart attack so that they can be pronounced brain-dead and their organs can be transplanted, *60 Minutes* does an expose and interviews a local district attorney who threatens to charge the surgeons with homicide if they persist.[16] Suppose physicians somewhere maintained the bodies of adult humans, whose higher brains had never been allowed to develop (at least the brain stem must be kept to regulate organ functions). This would mean excising the cerebrums of sentient human babies at birth. Even if such babies reached adulthood, how could their organs be used for others? There is no way this could be done secretly because organs must be matched by blood type, tissue typing, and size to recipients. Even if we could imagine it being done successfully in secret, the violation of medical ethics would be so horrendous that the expose would be the equivalent to discovering that infamous Nazi physician Sigmund Rascher was still torturing humans at Ravensbruck in his Sky Wagon.

Both Stephen Jay Gould and bioethicist Ruth Macklin have emphasized that the non-people argument is ridiculous because no one today ever considers treating one of a pair of twins as a non-person.[17] Someone may reply by saying that twins or triplets occur naturally, whereas cloned people do not. The premise assumed here is that human worth is maintained if genetic replication occurs randomly, but not if it were deliberately created. If it were deliberate, the resulting beings would be sub-human. So we reach the reductio conclusion that deliberate creation of a person lessens his worth.

I am going to insist henceforth on one reformative, linguistic stipulation to improve thinking about this topic. At this point in human history and with the legacy about human cloning we have from movies and popular fiction, to refer to people originated by cloning as "clones" or "a clone" is to drastically bias the discussion at the outset in the worst possible way. Popular discussions are full of question-begging phrases, e.g., "to maintain diversity in the human gene pool, it's important that there be no escaping clones" or "if we get some viable clones" or "a clonant . . . could rightly resent having been made a clone" (Leon Kass).[18]

"Clone" connotes sub-human, zombie-like, insect-like behavior. It is associated with phrases such as "an army of clones" and "Slave Clones of Caldor." It is really in the same class as dozens of other nasty terms that slur the racial, ethnic, and sexual origins of a person. So from here on, and despite the cumbersomeness of the phrase, I am going to write about "persons originated by cloning" or even more neutrally, about "nuclear somatic transfer" (NST) to produce a human or about "human asexual reproduction." For the same reason, I will avoid using "to clone" as a verb when it involves humans.

You Can't Reproduce Yourself

A lot of popular discussion about asexual human reproduction revolves around the question of whether a person could "clone himself." Indeed, most of this discussion assumes that it makes sense to talk this way.

Some of it is funny: "If you have sex with your husband's clone, are you really being unfaithful?" "Would a clone of Bhoutros Bhoutros Ghali's clone be named Bhoutros Bhoutros Bhoutros Ghali?" "If a cloned man and a cloned woman marry, divorce, change partners, and try again, it would be good," said a divorce lawyer. "The same reason the first marriage failed, the second marriage fails. That's four for me instead of two." Of course, to talk this way is to accept uncritically some very questionable assumptions about personal identity.

In Chapter 2, I discussed some bio-chemical reasons why one person originated from the genotype of another would not be exactly identical. Now I will discuss the environmental reasons.

Suppose I want to reproduce myself. Suppose I persuade my wife to undergo minor surgery, have an egg removed by laparoscopy, have its nucleus cut out and have my genes inserted. She will then spend nine months gestating the embryo, and because I might be killed in some accident, making an implicit commitment to raising the child until adulthood ("no small assumptions!'", says my wife in the background as I run this by her). Even if these assumptions came true, would the baby who was born be me?

Not likely. He would certainly not be an instant, carbon-copy of me. For one thing (and not to put too fine a point on it), he would be a *baby*, whereas I am 50 years old.

Moreover, he would grow up not in the years after the close of World War II in the suburbs of Washington, D.C., but in the suburbs of Alabama at the beginning of the 21st Century. He would not watch *The Mickey Mouse Club*, buy five-cent Cokes and return the bottle to get back a 2-cent deposit. He would not be the oldest of five children, with attendant, later baby-sitting responsibilities. Instead, he would grow up and know about MTV, know how to work a Macintosh computer and get on the Internet, know about the world from CNN, and have to adjust to dogs and cats in his house. More important, he would not have Gil and Louise Pence as his parents but Greg and Pat Pence, and the latter would be different parents than the former (not necessarily better, just different). So if I wanted to clone myself, I would be disappointed.

We do not know how much nurture contributes to personality over nature. (One thing that cloned humans would help teach us is the answer to this question; more on this in a later chapter.) But we believe nurture contributes a lot. Sometimes the contributions of nurture are masked as those of nature.

For example, scientists discovered in 1997 that how much a baby is talked to (and with how much affection) dramatically affects how many neural pathways are formed in its inchoate brain.[19] Any baby's brain is like a wild garden, where talking forms pathways that cannot later be formed. Once the garden grows too wild, it cannot later be tamed. By using NMR (nuclear magnetic resonance) scans of the brains of such children at age two, scientists were able to document huge gains or losses from the results of the verbal activities of parents. All of these pathways seem to get laid down by the age of two, and scientists found huge differences in how much they talked to their babies between (to use their categories) parents who were on welfare, who had blue collar jobs, and who were professionals.

So the interventions of parents may be much more important than genetic reductionism implies. The ultimate constitution of the phenotype, the observable physical traits of a specific adult, is in an ongoing, evolving interaction with a changeable, variable environment. And here the environment matters a great deal. A person originated from Hitler's genes could end up a rabbi. A person cloned from Michael Jordan's genes could be a televison weatherman who couldn't care less about basketball. A genotype originated from Placido Domingo might only be an average singer. Remember, milk from nine cows produced from nine twinned embryos is only predicted to be 70% similar.[20]

Finally, for a male to talk about cloning himself is really far-fetched. The only way he is going to do so is to find a willing woman to gestate the embryo with his replicated genes, and whoever that woman is, will be the gestational mother and contribute her mitochondrial DNA to the resulting child. Unless surrogacy is involved (not a likely prospect), that woman will also be the real, rearing mother, and hence, will contribute a lot to the development of that boy.

So if a man wants to replicate himself, he better think it through. It is likely that his relationship, or the lack of it, in the daily activities of raising a child will have as much to do with whether the child is like him than any connection he has to the child through his genes.

Will rich men hire surrogate mothers to gestate and raise their cloned embryos? Not too often. Only three American states currently legalize commercial surrogacy, i.e, they consider surrogacy contracts enforceable in court. Many states ban such contracts altogether. In any case, surrogacy is expensive and a lot of things can go wrong. A woman employed as a surrogate always has a legal right to abort. As the Baby M case taught us, the surrogate mother may bond with the baby and change her mind about giving up the baby.

However, I suppose that if a rich person were fanatical enough about originating a genetic copy of himself, he could attempt it. I don't see anything that would prevent him from flying to some offshore island with various fertility specialists. He could hire local women to gestate his genes in a re-implanted enucleated egg. He could give a huge bonus for actual birth, say five thousand U.S. dollars, encouraging any pregnant woman to take all possible precautions to avoid miscarriage.

Even then, under the law of whatever country he was in, the mother of the child would undoubtedly be the legal parent. Whether he could adopt the child would be up to the mother and the laws of that country, and it is certainly not a sure thing that all this would work out. The mother might bond to the child, change her mind, blackmail him for more money, or want to get married. Each woman who gestated his child might sue him in U.S. courts for child-support or other monies. Even if he adopted the child, he would need to wait many years, perhaps twenty, to see how closely the child resembled him. Since many rich men only get such fanatical ideas late in life as they see their death coming, it is unlikely that he would ever know if his experiment was successful.

Lack of Informed Consent Doesn't Matter

Another misconception can be dismissed very quickly. In his testimony, George Annas invoked an objection of conservative Christian fundamentalist Paul Ramsey, who argued in the early 1970s against all forms of in vitro fertilization because it constituted "non-therapeutic experimentation on the unborn without their consent."[21] After the announcement of Dolly's cloning, religion columnist Mike McManus echoed Ramsey's criticism, implying that because a child would never be able to give informed consent to being brought into existence through cloning, to do so would be unethical. Similarly, George Annas argued that because experimentation on humans without informed consent is fundamentally unethical, embryonic experimentation is therefore unethical because embryos can never consent to such experimentation: "The birth of a human from cloning might be technologically possible, but we could only discover this by unethically subjecting the planned child to the risk of serious genetic or physical injury, and subjecting a planned child to this type of risk could literally never be justified."[22]

The argument is so silly that it would not be worth mentioning but for the fact that otherwise sane people continue to use it. First, the requirement of informed consent to enter existence is preposterous. ("Charles, dear, please consent to be born. Look, it really isn't so bad out here. Let me tell you why. . . .") One has to be a person to give consent and one can't give consent without being a person, so one can't consent to become a person.

Second, if IVF or cloning is the only way that a child is going to exist, and if there is every expectation that the child will be normal (see below), then we can use something that researchers call presumed consent. This kind of consent is where we predict what reasonable people would consent to in experiments where actual consent is impossible (e.g., because some deception is needed that prevents consent). So here the question of presumed consent is: "Would you be willing to take a tiny risk at birth of defects in order to exist in the first place?" And the answer to that is obvious.

Scientists Aren't Frankensteins and Strangeloves

America's most-famous bioethicist, Art Caplan, saw the public's horror at human cloning not as a reaction to human cloning per se but as a symbol of the public's general distrust today of science and scientists.[23]

In 1818, novelist Mary Shelley wrote *Frankenstein, or the Modern Prometheus*. Remarkably, this nearly two-hundred-year-old novel has had a profound influence on Western culture, being the inspiration for hundreds of movies and books. Later we got *Dr. Jekyll and Mr. Hyde*, a cautionary tale about a physician who changes his nature (or perhaps the experiment merely allows his evil, twin nature to emerge?)

As he narrates his story to the reader, Dr. Frankenstein reveals that he could never control his rabid ambition to create an adult person, even when the welfare was at stake of others such as his wife Elizabeth. When he is successful, he is disgusted at the creature's crude features and qualities, and he immediately regrets his work, wishing his monster were dead.

The real monster in Shelley's view, of course, is Dr. Frankenstein, not his woeful creature. Shelley lets us hear the laments of the creature as he describes the prejudice and beatings he has endured.

In novels, there is redemption and transcendence. Because of the creature's ugliness and loneliness, he learns intimacy and love; Christ-like, he forgives his maker; human-like, he hates the mobs who persecute him. In the end, the creature is more human, more moral, than the scientist who created him.

This novel may have become such a classic because it tapped into our fear and distrust of scientists. We fear that some person may acquire a large bit of knowledge that can alter our fate. In 1964, *Dr. Strangelove* updated this fear, only instead of a monster from biology there was the much more deadly fear of nuclear war. Dr. Strangelove himself, in a memorable performance by actor Peter Sellers, seems modeled on German scientists such as Wernher von Braun, who were allowed to immigrate after World War II to help the American space program. But the loyalty of such men to their former enemy was not clear. Stanley Kubrick also makes it clear that Strangelove is at heart an eccentric little boy who is obsessed with scientific toys and who has fantasies of spending years in seclusion with "ten young women who need impregnation for the future of the race."

Such scientists are not people of moral wisdom, who parent children, and who have the humanizing influences of solid marriages. They lack compassion, common sense, insight into themselves, and maturity. Is it so surprising that modern culture does not trust scientists? Throw in the modern update from PETA (People for the Ethical Treatment of Animals)—of the scientist as animal-torturer—and the horrible picture is complete.

Dr. Strangelove in 1964 was based on the novel *Red Alert* by Peter George. Also released in 1964 was another novel that raised the nuclear theme, *Fail Safe*. All these novels raised the fear that science would create a new danger. This danger of the modern world occurs when a weird concatenation of events creates a situation where a disaster suddenly occurs. Science keeps its monsters in cages, but monsters are dangerous and can escape with one little slip.

Hence the great fear in the 1970s over the construction of special labs for recombinant DNA research in Cambridge, Massachusetts where some feared that ambitious scientists might create lethal viruses that would escape labs and kill the surrounding population. Conferences like the famous Asimolar conference led to the self-policing of scientists. Microbiologist Roy Curtiss created a fragile form of the E. coli bacteria to host gene-splicing experiments that would self-destruct if it got outside the lab. Stringent guidelines were enforced requiring that any potentially-dangerous experiment be done in highly controlled labs called "P-3 containment labs." Lost among the original worries was the fact that the first people to die from any such lethal virus would be the scientists themselves and their families.

Chalk up all the fears of that time to the influence of Michael Crichton's 1971 film *Andromeda Strain*, based on his novel of 1970. Again in that novel we see the theme of a few men, acting in secret with special knowledge, who willingly put all of humanity at risk because of their arrogance and stupidity. In this film, a crystalline life-form that can exist in space comes to earth when a satellite is returned and kills almost everyone in the small village where it lands. Put into a secret containment lab in the southwest United States, several predictable glitches occur, each of which almost dooms humanity: a piece of paper catches the bell of the teletype machine (so the bell doesn't ring and no one knows a message has been sent), and the President worries about politics and waits too long to drop a nuclear bomb on the village. Meanwhile, an automatic failsafe plan in the containment center involving nuclear destruction almost occurs, and is thwarted because it is later realized that the plan would actually grow the Andromeda Strain.

All of these movies have the same theme: scientists are clever but fools; they do dangerous things and risk their lives and our lives because of their fanatical pursuit of knowledge; knowledge is dangerous; science is dangerous. All have the same take-home message: science will soon release a genie from a bottle and not be able to put him back.

All of these movies and novels aim to entertain and to make money from our desire to be entertained. They do not aim at telling the truth. As such, they should not be the basis for thinking about creating a human by cloning.

George Annas also said that human cloning would be morally wrong because "it is the manufacture of a person made to order . . . and symbolizes the scientist's unrestrained quest for mastery over nature for the sake of knowledge, power, and profits."[24] Annas here labels all scientists Drs. Frankenstein, imputing to them an "unrestrained quest for mastery" over the world. Of course, "mastery" conjures up Hitler's "master race," as does "unrestrained."

In truth, scientists are restrained quite a lot by federal and moral constraints. Human experimentation since 1972 must be approved by local review committees (called "IRBs" for Institutional Review Boards) and, if federally-funded, through national peer-review panels. Even animal experimentation must be approved by Animal Use and Care committees.

Scientists don't always follow the maxim, "If we can do it, we ought to do it." When it became possible to know the sex of a fetus early in gestation, genetic counselors debated whether to tell parents the sex of a fetus when parents indicated they only wanted a child of a certain sex. The universal reaction by counselors was that such a desire was undesirable and to accede to it by counselors would be inappropriate, so voluntary guidelines were established that have been honored almost without exception.

Moreover, why even buy into the Frankenstein nightmare? The average life of the average scientist is hardly a "quest for mastery over nature for the sake of knowledge, power, and money." (And this is a lawyer making this charge!) At my university, there is a young, internationally-known animal geneticist working on producing a mouse that will be a model for Huntington's disease in humans. Pete Detloff is trying to produce a mouse that has the DNA repetitions that genetic research has discovered to be characteristic of Huntington's. He has a team of young post-doc's, pre-doc's, lab staff, assistants, and students who work with him, almost all of whom are under 40 and extremely dedicated. Near the deadline for grants, which fund this kind of work and that of most of Pete's colleagues, everyone on the team may work 80 hours a week. And the pay is far less than they would make in real estate or on Wall Street. Yet we do not fear those on Wall Street or in real estate, even though they can downsize our job or build a development that destroys our community.

I believe that over the last two decades, popular culture has turned against science. Science is no longer worshipped as the key to progress and it has become fashionable to question the objectivity of scientists at every turn: feminists lambast male values that supposedly control ways of knowing and valuing, and proponents of alternative medicine (many of them M.D.s selling books!) say traditional scientific methods don't bring results fast enough.

By far one of the greatest assaults has come from those who see all scientists as animal-abusers. Less than one percent of scientists engage in research on animals that most people would find objectionable, yet most scientists feel the need to defend animal research when it is simplistically attacked, and to circle the wagons with their fellows. Some of the proponents of animal rights have used cheap, emotional tricks to raise opposition and money, and have painted all scientists with the same brush in doing so.

This is so unfortunate. It is also so obvious that the real animal abusers are not mad scientists in distant laboratories but the neighbor down the block who doesn't responsibly care for a dog or cat. Of course, it is easier to blame strangers for the abuse of animals than the evil we see everyday.[25]

Reproductive Freedom Doesn't Lead to Coercive State Eugenics

There is an invidious association that different forms of originating humans lead to eugenics which in turn lead to state-controlled breeding and the destruction of reproductive liberty in ordinary couples.

Some of this is due to not understanding the point of the famous classic, *Brave New World*, by Aldous Huxley. Written in 1932, this novel has been often cited by social critics as having predicted the kind of results they now fear—a future in which governments use technology to control reproduction.

This view of *Brave New World* is mistaken. The controls Huxley had imagined were based on psychological conditioning and need to be seen in the context of behaviorism, a school of psychology which was then as feared and misunderstood as cloning seems to be now. It is mistaken to extend Huxley's fictional ideas about psychological manipulation to fears about real biological innovations

The great irony of all these citations is that in *Brave New World*, Huxley described the devastating consequences of loss of reproductive choice by individuals. Arguments that human cloning should be banned because of the message of *Brave New World* amount to saying that what the novel opposed should occur, that couples should be denied reproductive choices.

Conclusion

Some misleading ideas abound about creating a human by cloning. Many of them come from science fiction. Our history of false alarms about creating babies through in vitro fertilization should make us cautious about bans over any kind of human reproduction. Some of these misconceptions are so unreal as to be funny.

Notes

1. *Los Angeles Times*, 27 April 1997.
2. John Cardinal O'Connor, "Diminished Humanity: A Catholic View," *Special Edition on Human Cloning: REFLECTIONS: Newsletter of the Program for Ethics, Science and the Environment*, Department of Philosophy, Oregon State University, May 1997, 13–14.
3. John Locke, "Epistle to the Reader," *Treatise on the Human Understanding,*, Dover edition, 14.
4. C. J. Cherryth, *Cyteen* (New York: Warner Books, 1988), 6.
5. Carl Hartman, "Wilmut: Human Cloning Misguided," *Washington Post*, 15 April 1997.
6. Kate Wilhelm, *Where Late the Sweet Birds Sang* (New York: Harper & Row, 1974).
7. Fay Weldon, *The Cloning of Joanna May* (New York: Penguin Books, 1989).
8. David Brin, *Glory Season* (New York: Bantam Books, 1993).
9. George Annas, "Scientific Discoveries and Cloning: Challenges for Public Policy," Senate Subcommittee on Public Health and Safety, Committee on Labor and Human Resources, 12 March 1997. Testimony posted on the Internet at this address: http://www-busph.bu.edu/Depts/LW/Clonetest.htm.
10. George Annas, "Scientific Discoveries. . . ."
11. For Hollywood's treatment of twins, see: *The Dark Mirror* (1946); *Village of the Damned* (1960); *The Parent Trap* (1961); *Liquid Sky* (1981); *Dead Ringers* (1988); *Big Business* (1988); *Dominick and Eugene* (1988); *The Krays* (1991); *Double Impact* (1991); *Twins* (1988); *Steal Big, Steal Little* (1995). In literature, see: *Twelfth Night*, William Shakespeare; *The Man in the Iron Mask*, Alexandre Dumas; *Esau and Jacob*, Machado de Assiz; *The Prince and the Pauper* and *Pudd'n Head Wilson*, Mark Twain; Fall of the House of Usher, Edgar Allan Poe; *The Bobbsey Twins series; On the Black Hill*, Bruce Chatwin; *The Solid Mandala*, Patrick White; *Success*, Martin Amis; *Princess Daisy*, Judith Krantz; *Master of the Game*, Sidney Sheldon; and the *Sweet Valley High* series, Francine Pascal. (These lists courtesy of TWINSOURCE Internet site, where there is much more information about twins.)
12. Deborah Sharp and Lori Sharn, "Big Questions for Humanity," *USA Today*, 1 April 1997.
13. Leon Kass, "The Wisdom of Repugnance," *The New Republic*, 2 June 1997, 22.
14. Lori Andrews, moderator, "Legal, Regulatory, and Industry Issues," Conference on Mammalian Cloning: Implications for Science and Society, 27 June 1997, Crystal City Marriott, Crystal City, Virginia.
15. John Brown and Lesley Brown, *Our Miracle Called Louise* (London: Paddington, London, 1979), 98.
16. Gina Kolata, "Controversy Erupts Over Organ Removals," *New York Times*, 13 April 1997, 14.

17. Stephen Jay Gould, "Dolly's Fashion and Louis's Passion," *Natural History*, June 1997, 21; Ruth Macklin, "Splitting Embryos on the Slippery Slope," *Kennedy Institute of Ethics Journal*, 4, no. 3 (September 1994), 217.

18. Leon Kass, "The Wisdom. . . ."

19. Sandra Blakeslee, "Babies' Brains Are Dynamos of Intellect," *New York Times*, 17 April 1997, A1, A14; Sharon Begley, "How to Build a Baby's Brain," *Newsweek*, Spring/Summer 1997, Special Edition, 28–32.

20. Tim Friend, "Getting to the Nucleus of Cloning Concerns," Panel with Colin Stewart, Tom Murray, and Neil First, *USA Today*, 1 April 1997.

21. Paul Ramsey, *Fabricated Man: The Ethics of Genetic Control* (New Haven, Conn.: Yale University Press, 1970).

22. George Annas, "Scientific Discoveries . . . ."

23. Arthur Caplan, "Ethical Issues," Conference on Mammalian Cloning: Implications for Science and Society, 26 June 1997, Crystal City Marriott, Crystal City, Virginia.

24. George Annas, "Scientific Discoveries. . . ."

25. To its credit, PETA (People for the Ethical Treatment of Animals) many years ago began to take some responsibility for its previous focus on mad scientists. It broadened its focus to include ordinary people who buy puppies and then leave them as adults chained in the back yard while they watch ice skating on television.

Four Questions about Ethics

The first stage [of modern moral philosophy] is one of gradual emergence from the traditional assumption that morality must come from some authoritative source outside of human nature, into the belief that morality might arise from resources within human nature itself. It was a movement from the view that morality must be imposed on human beings towards the belief that morality could be understood as human self-governance or autonomy. This stage begins with the Essays *of Michel de Montaigne and culminates in the work of Kant, Reid, and Bentham.*

During the second stage, moral philosophy was largely preoccupied with the elaboration and defense of the view that we are individually self-governing, and with new objections and alternatives to it. The period extends from the assimilation of the works of Reid, Bentham, and Kant to the last third of the present century.

[In the last stage today], the attention of moral philosophers has begun to shift away from the problem of the autonomous individual toward new issues concerning public morality.

J. B. Schneewind, "Modern Moral Philosophy"[1]

· · · · ·

In this chapter, I describe four questions to ask when thinking about the morality of human asexual reproduction. Before these descriptions, it will be helpful to have a case for focus. (This case, although realistic, does not refer to an actual case.)

Case #1—Sarah and Abe Shapiro

Sarah and Abe Shapiro yearned for a child for years before being able to have one. Both came from large Jewish families that put great emphasis on parental involvement with children and on family activities such as playing sports, eating nightly meals, and going on long camping trips.

Sarah and Abe also inherited something else from their families. Tay-Sachs disease runs in Jewish families of Eastern European origin. It is a lethal genetic disease that produces children who always die before they become teenagers.

Knowing their risk, Sarah and Abe used in vitro fertilization (IVF) so that any embryo implanted in Sarah could be screened for Tay-Sachs. In IVF, three embryos are often implanted in hopes that one will successfully gestate.

Such was the way Michael was created. Unfortunately, when the embryo that would become Michael was moving down Sarah's fallopian tube, it damaged her tube and rendered her infertile (her other tube was already damaged). So the Shapiros resigned themselves to having one child of their own and hoped, perhaps, to adopt another later.

When Michael was four, he and Abe were driving home from an outing when a drunk driver smashed into their car, instantly killing Abe and rendering Michael comatose, but with a beating heart sustained on a respirator.

After Abe's funeral, Sarah hoped for Michael to recover, praying to God for a miracle, which unfortunately did not come. During this time, she mourns the death of both Michael and Abe. After a year, her rabbi and therapist urged her "to move on with your life." They want her to agree to remove the respirator and allow Michael's body to die. She is only 40.

Sarah does not want to remarry. She is a writer and now owns her own home because of Abe's life insurance. However, she misses having a child in the house.

At this point, she decides to have one of her eggs removed, its nucleus taken out, and have the genes from Michael's body inserted in her egg to create a new embryo. After doing so, she will let Michael go. One of her reasons for using Michael's genes, she says, is that, "I couldn't bear to have a child who then died very young of Tay-Sachs." In this way, she knows her child will also be normal and be part of both her and Abe.

In many sessions, the rabbi, therapist, and infertility-physicians explore with Sarah the idea that she is merely attempting to replace Michael and that she has not fully accepted Michael's death. These professionals want to ensure that Sarah understands that the new child will be very different from Michael. They emphasize that Abe's influence will be missing, that Sarah's egg will contribute mitochondrial genes, that Sarah herself is now different, and so on.

Sarah claims that she is not trying to mechanically replace Michael and that she has accepted Michael's real death. She adds, "I know Michael and Abe are dead, but if God lets me bring forth this new child, whom I will call David, then Michael's and Abe's lives will not have been for nothing, for in David's life I can see, if not them, then at least their features and talents live on. Maybe I'll see Michael's way of laughing and Abe's swagger after he performs well in sports. What's wrong with that?"

A genetic counselor points out that she may also get the worst qualities from Abe and Michael, and is she prepared for that? "The worst qualities?" she ponders. "Well, they sure weren't perfect and they did have some of those, but I personally would rather have their worst qualities than just accept some anonymous sperm implanted in me, where the child will have no relation to Michael or Abe, and perhaps, to a history of Jews going back five thousand years."

Query 1—Does the Rule Intrude Too Much on Personal Liberty?

John Stuart Mill wrote *On Liberty* in 1859, and it contains an admirable distinction between private life and public morality, a distinction based on the concept of harm. Mill believed that a civilized society must promote certain ideals and discourage certain vices. Society can do this through its public policy while granting individuals a sphere of private action that is protected from interference by government. Power of the nation-state can be dangerous when used against the individual, and so the agents of government—such as police and military—should be forbidden to meddle in private life.

Equally, Mill held, the majority of citizens should be forbidden from becoming tyrannical. It should be forbidden from imposing its religious beliefs on a dissenting minority, even indirectly—say, by a judge who insists on a Christian prayer with a jury before they hear a case. It should be forbidden from censuring what is discussed publicly, say by a television station that decides that its viewers should not see homosexual characters. It is important to emphasize here that Mill believed that the majority's tyranny is normally done in the name of morality.

It is natural to ask where the line is to be drawn between private and public life. Mill's rough rule-of-thumb is called his *harm principle:* private life encompasses those actions of adults that are purely personal and put other people at no risk of harm. In such areas, there should be no interference by government—even for a person's own good. Consider non-traditional sex roles between two consenting adults (where the wife leaves home to work and the husband raises the children at home): even if other people consider these roles immoral, their relationship for Mill poses no moral question because no one else is affected.

Building on Mill's work, we can distinguish between four different areas where issues about human cloning arise: (1) personal life, (2) morality, (3) public policy, and (4) the law. Issues of personal life are purely private and affect no one else. When someone else is affected, issues move from the personal to the realm of morality. When society attempts to promote certain positive values while at the same time tolerating personal disagreement with those values, we move into the third area, public policy.

Actions in the area of public policy, like those in the area of morality, do affect other people's interests, but persuasive actions in public policy are not necessarily condemned as immoral. Consider alcohol. Although society tries to discourage consumption of alcohol (by taxation) and regulates it (forbidding alcohol at elementary schools) people may drink in their homes without being viewed as immoral. Consider also adoption. Society wants adults to adopt needy children, and offers tax incentives to adults to so encourage this, but no one thinks it immoral for a childless couple not to adopt a baby.

These spheres overlap and shade into each other, and there is no exact criteria for separating one area from the next. The general goal is to limit the range of morality from two ends: first by carving out a zone of private, personal life, and second, by allowing society to encourage and discourage behaviors in public policy without explicit moral judgment or legal penalty. The general goal recognizes that we are all better off not moralizing every aspect of life.

One of the things Mill meant is that views that are essentially religious, even if held by the majority, should not be imposed on the minority. Especially in areas so personal as the make-up of the family and familial reproduction, the religious views of the majority have no place running federal policy.

Query 2—What Is the Point of the Moral Rule?

Instead of the usual question about ideal morality (about how morality ought to be), it is useful to consider how morality actually works. Call this the functional view of morality.

In this view, moral rules exist to adjudicate conflicts between the interests of persons. Because modern society contains many different kinds of people with many different points of view, moral rules are necessary for us to get along peacefully. In this functional view, the point of moral rules is not to prepare everyone for salvation or to create a purely religious state on earth. (These were the metaphysical beliefs associated with moral rules that at one time were quite functional.) Nor is the point of morality to create the greatest good for the greatest number of humans and animals on the planet. Nor is the point to create a perfectly rational, elegant theory of morality. Instead, the point is the more minimal one of getting along in a world where some resources will always be scarce, where interests of people conflict, and where people are interdependent and must cooperate.

So moral rules adjudicate social relations. Where they fail, the tougher ways of the law begin. Given that function, past moral rules may not always work in contemporary times, and when that happens, the nature of morality itself comes into question. For example, the very concept of having an interest has changed substantially over the last century, from covering one's household property to covering one's interest in pirated copies of one's book sold in China.

Moral rules in this functional sense are moot when there is no conflict, where the people have no real interests at stake, or where there are no existing people. For example, suppose Smith and I agree to share the planting of a boundary hedge along the property line on the west side of my yard, but my neighbor Jones on the eastern side is jealous of the cooperation between Smith and me, so much so that he objects to the joint project between Smith and me. Jones has no right to do so because he has no interests at stake. As so often happens in morality, his very objection creates a moral issue between Jones and me (because he is trying to interfere with my relations with my other neighbor) when there was no moral issue before.

Thus the point of moral rules is not to create an ideal society. Some philosophical vision of the future must do that, while moral rules allow us to get along enough to get there. In the technical language of moral philosophy, there is the theory of the right and the theory of the good. If we have the right theory of the right, we will allow different people to live their lives according to their view of the good.

Application of this point to human asexual reproduction is obvious: if there is no conflict between two or more people, there is no moral issue present. Despite the widespread belief to the contrary, if no one is harmed by human asexual reproduction, then it raises no moral issue.

I want to also make a more general point here about the point of moral rules. The two great traditions that we have inherited from the past focus on two ways of evaluating moral acts: by their motives and by their consequences. Hence, if we want to know why an action is right, we can look at either the motives of the agents or the action's consequences. Judaeo-Christian ethics tends to focus on the motive of the act—what was in the agent's mind or heart—and not on what consequences occur. A secular ethics such as utilitarianism focuses on the actual consequences.

This surprisingly simple fact—that motives and consequences determine the morality of an act—is a helpful one to keep in mind when we ask why a certain rule is still a good one. Sometimes, a rule will become written in stone and we forget why it came about in the first place. If we carefully inspect the motives and consequences associated with that rule, we may sometimes discover that it is outdated.

Nor should we assume that the specific moral judgments that we make and that seem "obvious" to us will stand the test of time. As the Australian moral philosopher Peter Singer writes on this question:

> Why should we not rather make the opposite assumption, that all the particular moral judgments we intuitively make are likely to derive from discarded religious systems, from warped views of sex and bodily functions, or from customs necessary for the survival of the group in social and economic circumstances that now lie in the distant past? In which case, it would be best to forget all about our particular moral judgments, and start again from as near as we can get to self-evident moral axioms.[2]

For example, our culture traditionally has forbidden actively assisting a terminally ill person to die ("active euthanasia"), but it considers it permissible to merely watch such a person die slowly. So a basic rule in our culture is that allowing terminal patients to die is permissible, where assisting them is not.

This rule is outdated. How do we know? Because in both modes, the motives and consequences are the same. Situations often arise with terminal patients where the motives of everyone—including the patient himself—are to create a quick, painless death. Here the people intend quick death and quick death is the result. Given such a situation, it cannot matter morally whether the actions taken to hasten death are passive or active.[3]

Put differently, if the motives were bad and the consequences were bad, then the action would be bad according to either kind of moral theory. But it would make little sense to say that the description of the act as passive or as active really held the moral weight. To say so would be like saying that a performance of a piece of music was bad not because of how it was played but because of how it was classified.

Query 3: Why Assume the Worst Motives?

The case of Sarah Shapiro is deliberately formulated to have a parent who has good motives about originating a child asexually. As the case shows, such a possibility is not unimaginable and, given the unpredictability of human life, a case such as this will one day arise.

Most popular discussions about cloning a human assume the worst possible motives in parents, but why on earth make such assumptions? Without evidence? If someone assumes that every person he meets is a secret racist or anti-Semite, we say he is paranoid, or a misanthrope, or warped. Why assume the worst motives when we are thinking about morality? Or in public policy? This way of thinking got us nowhere in the cold war, when the U.S. and Russia competed in the nuclear arms race and where it was assumed that Communists were evil people and Americans were saints. Why assume in public policy what we don't assume in ordinary life? We don't forsake participation in car pools that take our kids places because we fear that some parent may decide to kidnap the kids for ransom. Why should we assume worse when it comes to thinking about parents in public policy?

It has always been a trick of advocates of the status quo to assume the worst motives in humans. That is what the theory of original sin is all about. But humans are a lot better off today than a thousand years ago, and also a lot better off than a hundred years ago. And the main reason why is the electricity, antibiotics, clean water, efficient transportation, mass communication, and public education that humans have created. (Those who disagree know only the false, rose-colored versions of history seen in the mass media.) So why not trust humans rather than fear them? Who else has brought this progress? (If God has allowed humans to progress, why won't he allow them to progress more?)

An important corollary here is to ask about the evidence for assuming bad motives in ordinary people. If there is no such evidence, no such motives should be assumed. We have thousands of years of history with human parents and we know them well. We know that most parents most of the time do not have evil motives toward their children.

Nevertheless, many of our pundits assume the worst about us. Catholic University law professor Robert Destro wondered if cloned humans would have adequate legal rights "if they were created to perform specific work."[4] Why assume this? It is like saying that we should not admit emigrants to this country because they might by enslaved by natives. Why would a parent be so bigoted? ("Laura, dear, why don't we clone a little slave-child to walk the dog and clean the kitty litter?")

The Reverend Richard McCormick said that "the obvious motives for cloning a human were 'the very reasons you should not.' "[5] Obviously, Father McCormick thinks it is "obvious" that couples have bad motives. He thinks that a couple might try to "create someone who could be a compatible organ donor." Really? Create your son and rip out his heart?

McCormick was probably thinking of the Ayala case where a couple conceived a daughter as a possible donor of bone marrow for their elder daughter dying of leukemia, and where they were lucky and had a new baby whose marrow matched.[6] But as medical sociologist Jay Hughes notes, there is all the difference in the world between renewable resources for transplantation, such as bone marrow, skin, urine, hair, and blood, and non-renewable human resources, such as hearts.[7]

Bioethicist Thomas Murray, a member of the Bioethics Advisory Commission, said, "Why are we uneasy about cloning? We might be worried over the dangers of excessive control over human reproduction, about the dangers of unbounded human pride."[8] But why assume that a government ban on human cloning is also not "excessive control over human reproduction?" Why assume that "unbounded human pride" is why couples would originate children by cloning? Why is giving couples more control over baby-making—which they have lacked through 99.9% of human history—a bad thing?

Why make such ridiculous assumptions about the motives of ordinary couples yet to have children? Go to your local neighborhood meeting, Parents-Teacher Association night, or Kiwanis Club and ask yourself: are all those people the kind of people who have bad motives? To assume bad motives in a crack addict or an alcoholic parent is understandable because we know that their free will has been largely overtaken by a drug. The drug will win out over any motive for a child's welfare. But most parents are not drug-dependent, nor are they malignant narcissists. Indeed, when we are almost exclusively discussing parents who want and plan for a child, and have good resources to raise such a child, we have adverse selection into that subset of parents who are unlikely to have such bad motives.

Query 4—Why Fear Slippery Slopes?

One of the central objections to cloning a human concerns the idea of a slippery slope, perhaps the second most famous idea in ethics (behind the Golden Rule). True, it will be allowed, extraordinary circumstances may make it plausible in the Shapiro case to think about allowing human asexual reproduction, but if that case is allowed, then another similar case must be allowed, until we get to some really terrible scenarios.

For example, twenty years ago in the debate about in vitro fertilization, Leon Kass objected that:

At least one good humanitarian reason can be found to justify each step. The first step serves as a precedent for the second and the second for the third, not just technologically but also in moral argument. Perhaps a wise society would say to infertile couples: "We understand your sorrow, but it might be better not to go ahead and do this."[9]

The rough idea here is that if a small, benign change is allowed, it will inevitably lead to another, less benign change, and so on through a series of inevitable steps, until a point is reached where a very bad outcome is at hand. A corollary is that, once the first change is accepted, there is no easy way to stop until the last, bad point is reached. Hence, the inference is made, better not to change at all.

The slippery slope is, for better or worse, also a central idea in bioethics. Because bioethics has been at the forefront of change over the last decades, "slope predictions" have been common. Indeed, every time real social change occurs, it scares most people, and some moralists will predict that the sky will soon fall: "The dawn of the era of cloning is a little like splitting the atom," said Dr. Glenn Bucher, president of the Graduate Theological Union in Berkeley, California, "with enormous prospects for evil and enormous prospects for good."[10]

But we must not be manipulated by predictions made at the drop of a hat. In the one above, with what is Bucher comparing "enormous prospects for evil"? The Holocaust? The Mongol invasion of Europe? AIDS? Does he really mean to indirectly refer to the atomic bomb?

One famous book was full of slope predictions. Thirty years ago, Alvin Toffler breathlessly coined the term "future shock" to "describe the shattering stress and disorientation that we induce in individuals by subjecting them to too much change in too short a time."[11] His *Future Shock* sold millions of copies and he was anointed as the futurologist whose omniscience revealed the (mostly dire) future of humanity. Toffler hyperventilated that social change was occurring so fast that we were losing all our moorings and would soon be adrift in a sea of social chaos. (Alasdair McIntyre's books push the same theme at the theoretical level in ethics.[12])

Toffler wrote *Future Shock* between the years of 1965 and 1970, when the industrialized, Western world was rapidly changing. Those years witnessed big changes in music, sex roles, blended families, suspicion of authority and old age, and a new tolerance for drugs, sexual experimentation, contraception, abortion, and divorce.

What Toffler failed to predict was that too much change creates an opposing reaction toward stability. By 1981, when AIDS began, the conservative reaction was already well under way and it kept rolling through the 1990s: couples reverted to traditional sex roles, nuclear families were again seen as an ideal, hostility renewed towards illegal drugs (especially cocaine and heroin), realization occurred that contraception and abortion weren't stopping teenage pregnancy, and divorce was seen to hurt children and hence, to be too easy. If we slipped down the slope, and many would deny we did, then at some point we took stock of where we were, changed our minds, and walked back up.

The specific predictions made by *Future Shock* about human cloning, artificial wombs, and genetic engineering are lessons in caution. Nobel Laureate geneticist Joshua Lederberg predicted to Toffler—sometime between 1965 and 1970—that "somebody may be doing it [cloning] right now with mammals. It wouldn't surprise me if it comes out any day now."[13] As for cloning humans, Lederberg gave it (at most) fifteen years. Lederberg also thought that the time was "very near" when "the size of the brain . . . would be brought under direct developmental control," when we could create much bigger, better brains for children.

One of the great problems for a non-scientist in the field is to evaluate the ability of someone like Lederberg to make such predictions outside his real field of expertise. Lederberg sounded perspicacious at the time, and certainly exciting (and Toffler was certainly selling excitement about the future in his book), but Lederberg ignored countless barriers, such as the ability of the government—if it chose—to ban funding for such research.

And as for Toffler, of course it is the tone that sells a book, especially a tone of impending Armageddon:

> It is important for laymen to understand that Lederberg is by no means a lone worrier in the scientific community. His fears about the biological revolution are shared by many of his scientific colleagues. The ethical, moral, and political questions raised by the new biology simply boggle the mind. Who shall live and who shall die? What is man? Who shall control research into these fields? How shall new findings be applied? Might we not unleash horrors for which man is totally unprepared? In the opinion of many of the world's leading scientists the clock is ticking for a "biological Hiroshima."[14]

Well, not really. And I would like to see the hard data that proved, even then, that "many" of the world's top scientists feared such a future, or that Lederberg's views were not confined to a small, speculative minority. In fact, Lederberg was very alone in going out on a limb with his highly speculative predictions.

In the next paragraph, Toffler quotes E. Hafez (a man who, he tells us, is an "internationally respected biologist") who predicted in 1965 that,

. . . within a mere ten to fifteen years, a woman will be able to buy a tiny embryo, take it to her doctor, have it implanted in her uterus, carry it for nine months and then give birth to it as though it had been conceived in her own body.

It wasn't until 1978 that Louise Brown was born by in vitro fertilization and the first American IVF baby didn't come until 1980. Only in 1996 did some desperate, infertile couples start to pay young women for eggs that would be fertilized with the husband's sperm for implantation in the older woman. Couples still can't "buy" an embryo.

Toffler next quoted Daniele Petrucci (by the way, all his quotes from Hafez and Petrucci came from a sensationalistic article in *Life* magazine in 1965, so Toffler was taking *Life*'s word about the credentials of these men and women, who claimed that artificial wombs are just around the corner):

Indeed, it will be possible at some point to do away with the female uterus altogether. Babies will be conceived, nurtured and raised to maturity outside the human body. It is clearly only a matter of years before the work begun by Dr. Daniele Petrucci in Bologna . . . makes it possible for women to have babies without the discomfort of pregnancy.[15]

Petrucci had claimed to have fertilized a human egg in vitro, grown it for 29 days, and then destroyed it because it was growing as a monster. What Toffler didn't discover then was that the evidence for this claim was never provided by Petrucci and the claim was later dismissed as fraudulent. (This fraud was harmful because it fueled later worries that IVF might produce monstrous babies—a fear also raised about cloning.) And of course, we are nowhere near having a real artificial womb.

In (what we can now see as) a hilarious scenario, Toffler somberly quotes Hafez's suggestion that,

fertilized eggs might be useful in the colonization of planets. Instead of shipping adults to Mars, we could ship a shoebox full of such cells and grow them into an entire city-size population of humans. Dr. Hafez observes, ". . . why send full-grown men and women aboard space ships? Instead, why not ship tiny embryos, in the care of a competent biologist . . . We miniaturize other spacecraft components. Why not the passengers?"[16]

Of course, Toffler could not resist the standard, dire predictions about eugenics, about a super race, and about state-controlled genetic enhancement. He eagerly quotes a kooky Soviet biologist predicting a "genetic arms race" between the Cold War enemies. For Toffler, "we are hurtling toward the time when we are able to breed both super- and sub-races. . . . We will be able to create super-athletes, girls with super-mammaries. . . ."

All these predictions were presented not as science fiction but as factual predictions. Toffler certainly got a lot of attention, but is his legacy a good one? On the good side, he scared people, and made them realize a lot of change had occurred in a few years. On the other side, he also made people feel that the change was uncontrollable and that we could never go back. In those aspects, his legacy has not been a good one.

Other breathlessly-made predictions haven't come true. In the 1960s, computers were seen as the oppressive agents of the State, but in fact personal computers later created new ways of sharing ideas that helped bring down Communism all over the world. Physician-assisted dying for competent, terminal adults in Holland was predicted to turn that peaceful country into an ethical hell, but the practice has been going on for twenty-five years with hardly any bad results. Abortion has been legal in America for a similar twenty-five years and American society continues to function quite nicely.

All these changes—with computers, assisted reproduction, euthanasia, and abortion—were predicted by various seers to land us on an inexorable slide down the slippery slope. None of them came true. So the lesson here is easy: be wary of slope predictions and don't let them make you fear the changes that may bring you a better future.

Finally, one way that the first and last tests of this chapter are linked is that the slippery slope predictions often assume bad motives in parents. Ostensibly, desires to have children who lack genetic dysfunction and to make one's children as talented, healthy, and lovable as possible, do not seem like the pit at the bottom of a slippery slope—although from the way many pundits talk about the slippery slope, one might think it so.

Conclusion

I have offered four questions to ask when we discuss the ethics of human asexual reproduction. Of course, these tests are applicable to many other issues in ethics. In thinking about originating humans by cloning, we should not think of such origination as being a moral issue unless someone is harmed, not assume that traditional moral rules are always right because the problems they address may change, not assume the worst motives in parents, and not let predictions about slippery slopes make us fear change.

In the next chapter, I turn to a much more vague, but also pressing, question that human asexual reproduction raises in the minds of many, viz., the future of sex.

Notes

1. J. B. Schneewind, "Modern Moral Philosophy," in Peter Singer (ed.), *A Companion to Ethics* (Cambridge, Mass.: Blackwell, 1991), 147.

2. Quoted by James Rachels, in his *Can Ethics Provide Answers? and Other Essays in Moral Philosophy* (Lanham, Md.: Rowman & Littlefield, 1997), 8; from Peter Singer, "Sidgwick and Reflective Equilibrium," *Monist* 58 (1974): 516.

3. See James Rachels, "Active and Passive Euthanasia," *New England Journal of Medicine* 292 (9 January 1975), 78–80.

4. Gustav Niebuhr, "Cloned Sheep Stirs Debate on Its Use on Humans," *New York Times*, 1 March 1997.

5. Gustav Niebuhr, "Cloned Sheep . . . ."

6. See Gregory Pence, *Classic Cases in Medical Ethics*, 2nd ed. (New York: McGraw-Hill, 1995), 296.

7. Jay Hughes, Medical College of Wisconsin Medical Ethics listserv discussion, September 4, 1995. Transplanting a lobe of a liver or lung, or one kidney where two are functioning, is not transplanting a renewable resource but it is also not like transplanting a heart, which can only be done if a person is dead while his heart continues to beat.

8. "Overview on Cloning," *Los Angeles Times*, 27 April 1997.

9. *Newsweek*, 7 August, 1978, 71.

10. Gustav Niebuhr, "Cloned Sheep . . . ."

11. Alvin Toffler coined the term in 1965 in an article in *Horizon* magazine. The quotation is from his later book, *Future Shock* (New York: Bantam Books, 1970), 2.

12. Alasdair McIntyre, *After Virtue* (South Bend, Ind.: Indiana University Press, 1981).

13. Alvin Toffler, *Future Shock*, 198.

14. Alvin Toffler, *Future Shock*, 198.

15. Alvin Toffler, *Future Shock*, 199–200.

16. Alvin Toffler, *Future Shock*, 200.

Cloning and Sex

Be fruitful, and multiply.

Genesis, Chapter 9, 1

Attempts or hypotheses for obtaining a human being without any connection with sexuality through "twin fission," cloning, or parthogenesis are to be considered contrary to the moral law, since they are in opposition to the dignity both of human procreation and of the conjugal union.

Instruction on Respect for Human Life
(Donum Vitae), The Vatican, 1987.[1]

[Human] cloning would violate practically every sacramental dimension of marriage, family life, physical and spiritual nurture, and the integrity and dignity of the human person. In Orthodox [Christian] thought, many ethicists are ready to accept technological means to assist a husband and wife to conceive and bear children. We draw the line, however, at the introduction of a third party into that sacred relationship, for it transgresses the spiritual and physical unity of the spouses. How could we approve the substitution of a laboratory for one of the spouses?

Reverend Dr. Stanley S. Harakas
"To Clone or Not to Clone:
Orthodox Christian Views"[2]

· · · · ·

Given all the emotion around the debate on human cloning, there must be more to it than rational arguments for and against the possibility of originating humans by (to use the most neutral phrase) nuclear somatic transfer or "NST." For most people, the subject that immediately comes to mind concerns what NST means for sexual reproduction, and especially, what it symbolizes about human sexuality. Are we moving toward androgyny, where sexual attraction between members of two opposite sexes will seem primitive? Are we moving towards disvaluing human sexuality, such that sexual relations will come to be seen as something only crude people do? More pointedly, if some babies are created without sex, will that destroy the special nature of human sexual experience?

Making Babies without Sex: Morally Repugnant?

Leon Kass thinks it fitting that most people are repulsed by the idea of making babies without prior sexual relations.[3] He claims that the widespread moral repugnance against cloning correctly expresses a "horror" at how this process thwarts the process of our natural sexuality. For him, there has been a slippery slope from the sexual revolution (made possible by the extramarital use of the pill) to claims of gay men, lesbians, and single women for sexual rights, and then to various ways of assisted reproduction that involve making babies without sex (in vitro fertilization, NST). "Thanks to the sexual revolution, we are able to deny in practice, and increasingly in thought, the inherent procreative teleology of sexuality."

That last phrase has a pedigree, but one with interesting twists and reversals. For fifteen hundred years, Christian theology accepted the views of Augustine, a fourth-century philosopher and theologian who taught that the desire for intercourse ("concupiscence") was evil. Augustine's innovation over previous Christian admonitions to be virginal was to teach that marriage was the only context in which this desire could permissibly be fulfilled, and even then, only for the purpose of having children. For Augustine, having children within a marriage was a license to sin; and once a marriage had produced enough children, that license was revoked.[4] Augustine specified that original sin expressed itself in lust, and that sin was transferred through intercourse from generation to generation; Christian doctrine thereafter followed the views of this patriarch for well over a thousand years.

So if Augustine thought it imperative for Christians to reproduce themselves but continued to believe that original sin was transmitted in the sex act, he might have welcomed human asexual reproduction. With it, a man might never sin sexually in his life. He could allow a NST embryo to be implanted in his wife for each child desired. The marriage need never be consummated. In contrast to in vitro fertilization, even masturbation by the male would not be required (sperm could be removed surgically). The descent of original sin to a new generation might thus be stopped.

Move forward now to 1978, the year of Louise Brown's birth by in vitro fertilization. The Roman Catholic Church prides itself on its apostolic succession and its teachings of the traditional ideas of the church fathers. Yet it is important to understand that sometimes it completely reverses itself.

What was the recent theological reason for condemning in vitro fertilization? The reason was that no act of human sex had occurred to create Louise Brown. From being condemned by Augustine for carrying original sex, Catholic teaching now promotes the idea that the sex act has become an essential requirement for moral conception of a human being. After nine years of study, the Vatican *Instructions* of 1987 condemned in vitro fertilization, equating it with "domination" and "manipulation of nature."[5] One bishop said: "The Christian morality has insisted on the importance of protecting the process by which human life is transmitted. The fact that science now has the ability to alter this process significantly does not mean that, morally speaking, it has the right to do so."[6] The official position of the Vatican is that sexual intercourse between husband and wife is necessary for moral conception; therefore, in vitro fertilization and NST are condemned because they take place without intercourse.

Sex has thus been reformatted in Catholic theology, undergoing a conceptual shift. Sexual desire is no longer an evil force that lures young people to hell—it has now become an uplifting expression of love in happy marriages. What is so surprising is the frequent assertion of the continuity of Catholic teaching on sex, whereas in fact the new teaching contradicts Augustine's teaching. (And his teaching in turn contradicted the former teaching of Jesus and Paul that virginity was the ideal. Note that Paul's statement that it was "better to marry than burn" should be taken in context, and in the context in which he said it, virginity was definitely the ideal.)

Ancient Judaism saw human sexuality as good, but only in the marriage bed. The Jewish attitude originated in contrast to the ancient Greek, exploitive attitude to sexuality. Speculating on why ancient Judaism adopted its family-based attitude, the modern Jewish writer Dennis Prager argues that:

> Man's nature, undisciplined by values, will allow sex to dominate his life and the life of society. When Judaism demanded that all sexual activity be channeled into marriage, it changed the world. . . . This revolution consisted of forcing the sexual genie into the marital bottle. It ensured that sex no longer dominated society, heightened male-female love and sexuality (and thereby almost alone created the possibility of love and eroticism within marriage), and began the arduous task of elevating the status of women.

It is probably impossible for us who live thousands of years after

Judaism began this process to perceive the extent to which sex can dominate, and has dominated, life. Throughout the ancient world, and up to the recent past in many parts of the world, sexuality infused virtually all of society. Human sexuality, especially male sexuality, is polymorphous, or utterly wild (far more so than animal sexuality). . . .[7]

Many scholars would object to some of Prager's sweeping conclusions, e.g., Prager fails to emphasize the power imbalance inherent in Greek sexuality or that historical Judaism made the man powerful over the woman in the marriage bed. Nevertheless, there is some truth in what Prager asserts. Within the marriage bed, no real restrictions were placed on sex; and sex there was celebrated and encouraged. Moreover, Judaism did indeed steer sexuality away from homosexuality, prostitution, premarital sex, and extramarital sex.

In contrast, Alexander the Great's Hellenistic world had a darker view of sexuality. The Greek emphasis on this life and the perfection of the body encouraged a freewheeling attitude towards sex. Greek culture here was famously homoerotic and many famous Greeks preferred anal intercourse with young boys over sex with women.

Definitive studies by scholars such as Dover, Greenberg, Sussman, and Boswell paint a picture of wild, polymorphous sexuality in Hellenistic times, with adult Greek males having sex with both females and young boys.[8] As the classicist Martha Nussbaum emphasizes, the important aspect of such sexuality was not the gender of the sex partner but who penetrated whom.[9] Powerful people had the right to penetrate; the conquered, vulnerable, and weak were penetrated. As such, sexual relations could be said to be equally about sexual satisfaction and about power.

Many of these sexual relations were not consensual. Slavery was ubiquitous in Hellenistic times and slaves were at the sexual mercy of their masters. Prostitution of all kinds was equally rampant and simply accepted as a dominant, but ordinary, fact of life. Some religious temples, especially those of fertility goddesses, were practically brothels. Children had no legal or ethical protection whatsoever other than "natural" parental affection, and it was not uncommon for a parent to sell a young child of eight into lifelong slavery or prostitution.

So it is not surprising that ancient Judaism saw human sexuality as a powerful and dangerous force that must be contained by rules. One very important ethical rule was the one that channeled this force to relations between husbands and wives.

Thus ancient Judaism and ancient Christianity did agree on one thing:

human sexuality was a dangerous force. A major task of civilization was to control, educate, and channel this dangerous desire into the appropriate direction.

So now we come to the conclusion of this section, and that is that there is an inherent dilemma in many contemporary views about human sexuality. On one hand, if human sexuality is good in the marriage bed (Judaism), permissible in creating children (Augustine), good in expressing love in marriage (modern Catholicism), and if human sexual relations are pleasurable to the people involved, how can it be argued that human sexuality is not good in all its forms where no one is harmed? Why limit a good thing to only one sphere? This is precisely the reasoning of unmarried teenagers having sex, adults who live together and who have sex without being married, and gay men and lesbians, who have sex without the intention of having children or the ability to be legally married (though some do).

On the other hand, if human sexuality is dangerous, then what is so good about it that makes the lack of it an evil in asexual reproduction? There seems to be a lot of evidence that, for many people, human sexuality is not an easily controllable power. "Two-thirds of American women aged 15–44 are at risk of unintended pregnancy," says the Alan Guttmacher Institute, the definitive source for reproductive statistics about American women.[10] Witness the million American teenage girls who got pregnant after widespread availability of contraception in the early 1970s. That figure of a million pregnant girls per year has held constant for every year since then over the last two decades.[11] In the representative year of 1987, over 1.5 million women in America had births resulting from unintended pregnancies.[12]

The problem raised by this dilemma—is human sexuality inherently good or inherently bad?—is seen in Leon Kass' statement that:

[Human] cloning shows itself to be a major alteration, indeed, a major violation, of our given nature as embodied, engendered, and engendering beings—and of the social relations built on this natural ground. Once this perspective is recognized, the ethical judgment on cloning can no longer be reduced to a matter of motives and intentions, rights and freedoms, benefits and harms, or even means and ends. It must be regarded primarily as a matter of meaning: Is cloning a fulfillment of human begetting and belonging? or is cloning rather, as I contend, their pollution and perversion?[13]

But which "given nature" with respect to sexuality should we assume? That of Augustine or Kass?

For Kass, and with the possibility of cloning on the horizon, all the old worries about the dangerousness of heterosexual relations seem trivial and innocent. Now heterosexual relations are the foundation of all human society and civilization (our fundamental "social relations"). Now those who wish that nothing will change will elevate sex to lofty heights, using words that convey mysterious and profound connotations:

> . . . [T]he ontological meaning of [human] sexual reproduction . . . [is that] only sexual animals can seek and find a complementary other with whom to pursue a goal that transcends their own existence. For a sexual being, the world is no longer an indifferent and largely homogeneous otherness, in part edible and in part dangerous. It also contains some very special and related and complementary beings of the same kind but of the opposite sex, toward whom one reaches out with special interest and intensity. . . . In higher birds and mammals, the outward gaze keeps a lookout not only for food and predators, but also for prospective mates; the beholding of the many splendored world is suffused with desire for union, the animal antecedent of human eros and the germ of sociality.[14]

But the contrary view is also contained in this quotation, for sexuality is also something that humans share with the lowly non-human animals, and that part of sexuality has often been called our "animalistic" nature, our "bestial" desires, our lust.

Both sides can't be true. We can't share sexuality with the non-human mammals, think ourselves much better than such animals (clone them, not humans!) and then argue that our sexuality makes our lives special and "transcendently" meaningful. Which view should guide public policy: human sex as great or human sex as evil? Human sex as a source of meaning or human sex as a way to lose salvation? (Maybe neither: maybe we should just leave it up to individuals.)

Plain Sex

One might think that humans by now would possess a deep, common, conceptual understanding of human sexuality because the desire for it is so pervasive among humans and because our society emphasizes it so much. Yet some public intellectuals have very strange views of it.

One famous extreme view of sex came from Freud, who argued that all human desire is some form of sexual desire. In reacting against the denial of sexuality of Victorian culture, Freud saw repressed sexual desire everywhere. But he went too far and failed to see that in a sexually-open society, other desires than sex could take over.

On the other extreme from Freud is the kind of view that has dominated Western thought since Augustine. On this view, there is some purpose connected to sex that redeems it. Here, permissible sex is a means to fulfillment of this purpose. On this "means-ends" view, sex is good because of its connection to a purpose such as having children, embodying love, stabilizing marriages, enabling communication, or increasing awareness.

In a famous article, University of Miami philosopher Alan Goldman argued in "Plain Sex" that there is a reasonable view of sex that is an alternative to the above view and that also both demystifies and de-intellectualizes it.[15] Goldman's view is a useful antidote to the usual disease of making lofty metaphysical pronouncements about sex.

For Goldman, sexual desire is simply "the desire for contact with another person's body and for the pleasure which such contact produces; sexual activity is activity which tends to fulfill such desire of the agent." Aristotle thought that pleasure is normally the byproduct of an action (e.g., playing tennis) and that to seek pleasure as the goal in such an activity is the wrong way to go about being fulfilled. Goldman believes that sexual desire differs because the desire for the particular kind of pleasure is precisely what is sought.

Of course, one can quibble with Goldman's view and point out that the physical intimacy involved in sex may be more important to some people than its actual physical pleasures. Nevertheless, it is hard to argue with his overall point that the best way to understand the value and worth of sex is as "plain sex," not as a means to some other goal.

Social conservatives, theologians, and most clergy have historically been suspicious of any attempt to separate the pleasures of sex from something that redeems these pleasures. Origination of humans by cloning symbolizes this separation in an important way because there will be human beings whose very existence is witness to the new fact that sex and reproduction need not always go together. More important, these new humans will be walking symbols of the fact that we need not think of sex anymore as being connected to having children.

Meilaender's Testimony

When asked to testify about human cloning before NBAC, Lutheran theologian Gilbert Meilaender condemned human cloning on grounds that, he said, were distinctively theological and Protestant.[16] In doing so, he sharpened the contrast between Goldman's view and most theological views.

For Meilaender, the act of sex "is not simply a personal project undertaken to satisfy one's own needs." Sex cannot be understood simply as the desire for pleasure in another's body.

What if we choose otherwise? Well, Meilaender replies, "The meaning of what we do . . . is not determined simply by our desire or will." Here is fatalism working its ugly way through theological thinking. Lest we have any doubt of this, Meilaender's closing sentence urges Commission members to "remember that 'progress' is always an optional goal in which nothing of the sacred inheres." (Really? Progress and the sacred are incompatible? A lot of progress has occurred since the time of Jesus, so does that mean . . . ?)

Where does a theologian go to condemn human cloning? Where Augustine himself went in the fourth century when he was looking for a way to condemn sex as evil, especially sex outside marriage. Augustine focused on the passage in Exodus where Onan spills his seed rather than consummate Levirate sexual relations, after which God slays him (in Levirate marriage, the brother-in-law of a woman who becomes a widow without having produced a child is obligated to impregnate her to produce an heir). From this, Augustine made the notoriously bad inference that the passage indicated not disobedience to God's will (in not completing reproduction) but because God was offended at seeing Onan's "seed."

Meilaender goes back to where everyone goes, to the 5,000-year-old story in the first chapter of Genesis, which, he says, "depicts the creation of humankind as male and female, sexually differentiated and enjoined by God's grace to sustain human life through procreation."

Talk about selective reading! The problem with Protestants justifying their views on biblical passages is that they only go there to justify what they already believe, not to find guidance. Of course, in the other story of creation there is only one human, Adam, and it is only after God perceives him to be lonely does he make Eve out of Adam's rib.

Modern Christians invert the teaching of patriarchs such as Augustine and Jerome, because they have come to accept the secular teaching of everyone else that sexual expression in marriage is not evil. Still, Christians require it to be redeemed by the "end" of love. It is still no good if two people are not in love and just enjoy sex.

Over and over again we hear that Christianity asserts a "transcendent" and "mysterious" connection between sex and producing children. Meilaender testifies that,

The possibility of human cloning is striking because it breaks the connection so emphatically. It aims directly at the heart of the mystery that is a child. Part of the mystery here is that we will always be hard-pressed to explain why the connection to sexual differentiation and procreation should not be broken.

Talk about turning a negative into a positive. If the connection between procreation and two parents can't be rationally justified, well, then . . . just turn it into a mystery.

Meilaender wraps his fatalism in romanticized language. Men and women should not even see a child as the result of sex and their desire to conceive, but instead, "The child is therefore always a gift—one like them who springs from their embrace, not a being whom they have made and whose destiny they determine."[17] Normal sexual reproduction is a "surrender to the mystery of the genetic lottery which is the mystery of the child who replicates neither father nor mother but incarnates their union."[18]

If every pregnancy from sex is a gift from God, why not accept every "gift" that comes along? Why not ban abortions? Contraception? By taking human desire and choice out of the picture, we are left with passive acceptance of fatalism. (Half of one million teenage girls: "I was meant to be pregnant. It is God's gift to me.")

As for the "mystery of the genetic lottery," I'm sure it's quite mysterious to the child with fragile-X syndrome, Turner's syndrome, or a Cyclops baby. When does mystery become tragedy? When are we allowed to choose to have better babies? Never? When are we allowed to say to the Giver of the gift, "Gee, couldn't you do any better than that?"

Conclusion

Leon Kass thinks human cloning morally repugnant because it is the reductio ad absurdum of changes that allow: women to have contraception and abortions; couples to have in vitro fertilization, egg transfer for use surrogates; and gay men and lesbians to have rights. Meilaender also echoes some of these themes.

Of course, one way to block a slippery slope is to deny that the bottom of the slope is bad: the "pit" at the bottom of the slope may look pretty good to some. Personally, I don't think a world where the above changes had not been made would be a better world than ours: a world where the teenage girls who get pregnant in America each year would be made to become mothers; a world where the 20% of married women who have abortions would be forced to bring an unwanted child into the family; a world where gays and lesbians can still be fired or evicted simply for being attracted to a member of the same sex (as they can be, in most cities and states in North America).

Western thinking has gone from making sex an evil symbolic of Original Sin to making it a symbol of love, communication, and good marriage. As Goldman argues, neither is correct. Sex is sex, and need not be anything more.

Human cloning separates sex from reproduction, and hence, symbolizes—in fact and concept—this break. As such, it resonates in our culture in more emotional ways than simply as a new way of making babies or as a new way of helping infertile couples. It appears that the separation of sex from reproduction strikes some deep fears in the human breast, making theological pundits predict that we are fast approaching reproductive Armageddon. To all this, one thinks of Joan Rivers laughing, shaking her head, and exclaiming, "Oh, puh-leeze!"

Notes

1. Congregation for the Doctrine of the Faith, *Instruction on Respect for Human Life in Its Origin and on the Dignity of Procreation*, 1987. Quoted in NBAC, *Cloning Human Beings: Report and Recommendations of the National Bioethics Advisory Commission*, Rockville, Md., June 1997, 52.

2. Special Edition on "Human Cloning," *REFLECTIONS: Newsletter of the Program for Ethics, Science and the Environment*, Department of Philosophy, Oregon State University, May 1997, 3.

3. Leon Kass, "The Wisdom of Repugnance," *The New Republic*, 2 June 1997, 17–26.

4. Paul Johnson, *A History of Christianity* (New York: Atheneum, 1983); Elaine Pagels, *Adam, Eve, and the Serpent* (New York: Random House, 1988).

5. "Text of Vatican's Statement on Human Reproduction," *New York Times*, 11 March 1987, pp. 10ff.

6. Bishop Kelly, quoted in G. Vecsey, "Religious Leaders Differ on Implant," *New York Times*, 27 July 1978, p. A16.

7. Dennis Prager, "Judaism, Homosexuality, and Civilization," *Ultimate Issues* 6, no. 2 (April–June 1990), 2.

8. David Greenberg, *The Construction of Homosexuality* (Chicago: University of Chicago Press, 1988); K. J. Dover, *Greek Homosexuality* (Cambridge, Mass.: Harvard University Press, 1978, 1989); Norman Sussman, "Sex and Sexuality in History," in *The Sexual Experience*, Sadock, Kaplan and Freedman (eds.), (Williams & Wilkins, 1976); John Boswell, *Christianity, Social Tolerance, and Homosexuality* (Chicago, Ill.: University of Chicago Press, 1980).

9. Martha Nussbaum, "The Bondage and Freedom of Eros," *Times Literary Supplement* (London), 1–7 June 1990.

10. Alan Guttmacher Institute, *Teenage Sexual and Reproductive Behavior: Facts in Brief*, 1993 (New York: 111 Fifth Avenue, New York, NY 10003), 19.

11. Alan Guttmacher Institute, *Teenage Sexual* . . . , 22.

12. Alan Guttmacher Institute, *Teenage Sexual* . . . , 22.

13. Leon Kass, "The Wisdom . . . ," 20–21.

14. Leon Kass, "The Wisdom . . . ," 21–22.

15. Alan Goldman, "Plain Sex," *Philosophy and Public Affairs* 6, no. 3 (Spring 1977), 267–287.

16. Gilbert Meilaender, "Begetting and Cloning," *First Things* 74 (June/July 1997), 41–43.

17. Gilbert Meilaender, "Begetting . . . ," 42.

18. Gilbert Meilaender, "Begetting . . . ," 43.

ant in the uterus under normal conditions. Reductio ad
e embryos are in fact people with lost souls, why not bap-
them? Arrange funerals?[5]

mbryos are not persons because they fail to meet *the cogni-
onhood.* This criterion was (to my knowledge) first champi-
medical ethics by the late Joseph Fletcher.[6] On it, to be a
ble to think, to remember one's life, to be capable of
separates a normal, adult person from, say, a rat, is cer-
or reasoning, reflective self-awareness, communication,
d action), and consciousness of the external world. The
m is not that any one of these capacities alone is sufficient
but rather, that these capacities together define the core
up. The most important point is that a being lacking all
s, such as a human embryo, does not meet the cognitive
nce cannot be a person.[7]

here is quite simple: consciousness is the foundation of
without any conscious being, however beautiful it might
being were to see it, is not preferable to a plain world
he conscious being. Consciousness is the condition of all
, and agency.

criterion has a nice symmetry for both ends of life. It
ral years of irreversible persistent vegetative state is the
person. Both Karen Quinlan and Nancy Cruzan were long
r cases dragged through the courts.

rag us into the mire here of discussions of "potential per-
ng out that human embryos are potentially persons. This
easily dismissed in two ways. First, it drastically underval-
smisses the choice and commitment needed over nine
on and twenty years of child-raising by the mother to pro-
ult. It is like saying an acorn is a twenty-year-old, 60-foot-
lue and concept. But if I pick up an acorn from your yard,
e me with theft, but you would if I cut down your mature
a woman who has just gotten a positive result on a preg-
a double homicide charge. Even when the Old Testament
an unmarried woman is found pregnant, and she is to be
to death (Deuteronomy 22:21, Leviticus 21:9, Genesis
ever considers such a burning or stoning to be killing a
ndeed, embryos were not considered "persons" until very
y, around 1879.)[8]

f human embryos might be persons if willing women ges-
does not show that women have any obligation to do so.

Twinning Human Embryos

We believe that life begins at the beginning, not in the middle, and that even a single-cell child has the dignity of all the children of God.

John Cavanaugh-O'Keefe, Director,
Bioethics, American Life League, 1997[1]

[The separation of cells in a human blastomere to create multiple, genetically identical embryos is] intrinsically perverse.

Vatican editorial, quoted in NBAC's *Report*[2]

Let me suggest a conceptual approach that might be adopted. In the light of medical proposals to redefine death in terms of irreversible coma or a loss of the higher brain function . . . if such a non-cerebral or decorticate patient is no longer alive in any human sense or personal sense, would it not follow that a pre-cerebral embryo or fetus is not yet alive in any human and personal sense? This would, of course, obviate any further use of such question-begging rhetoric as "killing unborn babies."

Joseph Fletcher, *New England Journal of Medicine*, 1971[3]

· · · · ·

Overview of the Embryo Controversy

One might think that it would be a red herring to review the debate about whether dot-sized human embryos are persons with a right to life. Alas, that is not so. The issue of embryonic personhood arises in several contexts of real importance to human cloning. Federal funds have been under a de facto ban from funding research on embryos since the mid-1970s, and that ban shows little signs of lifting today. President Clinton imposed a similar ban in March 1997 for federal funds used to originate a human by nuclear somatic transfer. Will his new ban still be around in 2017?

It is important to understand the controversy over human embryos be-cause it has become linked to creating humans by cloning. It is also important to understand, as I shall argue, what we are losing in continuing the ban on experimentation on human embryos.

Many medically-beneficial uses of human cloning would involve twinned human embryos. The most well-known is that some medical centers have begun allowing couples to select embryos by sex in order to exclude those embryos vulnerable to sex-linked genetic disorders. The most exciting possibility here is single-cell diagnosis, in which one cell of an 8-cell embryo is removed and the DNA is replicated to provide material for genetic testing. Removal of this one cell does not harm the embryo. In 1994, this technique was first used to avoid creating a child with Tay-Sachs disease.[4]

Because of the research of scientists at the Roslin Institute, many new ideas about the uses of embryos are now floating around. The key discovery is that growing, embryonic cells can be made dormant (quiescent) by denying them nutrients. While dormant, it looks like it will be possible to subtract or add the particular sequences of DNA that compose a gene. Why is that useful for humans?

First, creating animals with desirable properties can now be vastly improved. Instead of breeding a desirable male with a desirable female and hoping that the random assortment of genes turns out well (which it will not, in most cases), a particular gene can be inserted over and over again in the same host female. Cloning enters in recreating the same host female, as well as the same embryo into which a particular gene is inserted.

Second, such techniques may allow the much-hoped-for development of transplantable organs from non-human animals. Previously, xenografts have been tried from baboons (in Baby Fae, from an infant baboon heart), other primates, and some—such as surgeon Christiaan Barnard—have thought pigs would also be good sources. All such transplants have been rejected by the human immune system, but if human genes were inserted into the pigs to make their hearts more human-like, the resulting hearts might transplant better. Specific proteins might be deleted that signal the human immune system to reject porcine organs.

Such transgenic animals will up the ante in the debate over using intelligent, non-human mammals to benefit sick humans. At the same time, the debate may move from debating theoretical benefits to debating immediate benefits to save identifiable human lives. Cloning may also actually reduce the number of animals used in research. Because hit-or-miss techniques can be reduced, the sooner we get a reproducible model in animals of each human disease, the fewer animals need be wasted.

If we just look at embryo research, it is a no-brainer that for some genetic disease, having dozens of copies of the same human embryo allows a scientist to perform experiments using some embryos as controls and the others

as the objects of experiments. W
new research on human developr

Of direct benefit to humans,
opened about how to "turn off"
cialized cells in human developm
the undifferentiated embryo prod
no longer "totipotent," capable c
cause we now believe that is false, i
to grow a new liver lobe or to gro
many kinds of cancer is as diseas
and growth. Through research on
well be possible to learn how to at l
Understanding that process migh
cancerous process.

One of the most exciting possil
trying immediately, is helping peop
ing a sheep model of CF. Cloned sh
therapy against CF. Another possibi
late re-growth of nerve cells after d
such as actor Christopher Reeve.

The point is that we won't know
are allowed to do so. Everything in
sheep or frogs. Human embryos ar
what goes on with them until we are
useful research with embryos, societ
the same as murdering humans.

Why Embryos /

The argument over whether or not pe
place most commonly in the context
of whether human embryos are perso
abortion, but opponents of abortion
justification: if society allows human e
human fetuses to be destroyed, and t
son" to mean that a being can be a "F
Fear of the slippery slope most plau
that personhood begins with the ferti
at conception. Such a view implies tha
prevent implantation of zygotes, kill '

bryos fail to imp
absurdum: if the
tize them? Nam

I believe that
tive criterion of pe
oned in moder
person is to be
cognition. Wha
tain capacities–
agency (motiva
most minimal c
for personhood
criterion as a g
of these capaci
criterion, and

The basic id
all value. A wo
be if a conscic
that contained
reflection, cho

The cognit
explains why
real death of t
since dead as

Some woul
sonhood," pc
argument can
ues and even
months of ge
duce a huma
tall oak tree i
you do not c
oak tree. Kil
nancy test is
relates that w
burnt or stc
38:24), that
second perso
late in Chris

Second, e
tated them,

To think otherwise has many reductio implications. Among others, it would seem to commit each of us, on an overpopulated planet, to "actualizing" as many potential persons as possible.

For these reasons, I do not believe that human embryos are persons; and so I do not believe that they should be treated as such. I believe they become persons by degrees over a continuum, such that it makes sense to think that an eight-month-old fetus is almost a person but an eight-cell embryo is not. The challenge is now to find a way to fund research on human embryos to help those of us who are persons.

Controversies about Embryos

In 1979, obstetricians Howard and Georgeanna Jones had established the first American IVF clinic at Eastern Virginia Medical School (EVMS). In October of that year, opponents of IVF jammed the auditorium of the Norfolk Public Health Department for a debate on the proposed program at EVMS, charging that it would inevitably lead to destruction of "tiny human beings" (these protesters also envisioned "mass production of artificially designed humans").

EVMS is a private institution, not subject to federal regulations, and so its program went forward. But the brouhaha resulted in a ban on federal funds for experimentation on embryos. Since ninety percent of experimentation in the United States is federally funded, this ban effectively stopped most American research on assisted reproduction. What good did this ban accomplish? Did it stop research on embryos?

Only in America. In other parts of the world such as England and Australia, research on embryos is allowed up to 14 days. As a result, Australian infertility companies now license breakthroughs to American physicians. And some infertile couples weren't able to conceive babies who might otherwise have been able to do so.

The Futility of the Human Embryo Research Panel

During the mid-1970s, Congress avoided facing the controversial issues of research on human embryos by delegating approval to a non-Congressional committee. This is a typical ploy by Congress when its members want to avoid taking the heat of voting on a controversial issue. The Ethics Advisory Board (EAB) was created and charged with approving any such research paid for with federal funds. The EAB eventually concluded that some such research was morally permissible.

Before its conclusions could be enacted, the Reagan administration came to Washington, D.C. Believing that destruction of embryos was linked to destruction of fetuses in abortion, the pro-life Reagan and Bush administrations did not renew the EAB's charter, which ended in 1981. Hence, no human embryo research could be approved.

As described in Chapter 3, a false controversy erupted in the fall of 1993 over a *New York Times* story that Jerry L. Hall had cloned human embryos when he had merely twinned them, putting human embryos and cloning back in the news (and, we might add, back during a relatively dull time in the media's relation to bioethics, at least compared to artificial hearts and AIDS). After all the hype died down, after the egg was wiped off certain faces, and after everyone realized that no humans had really been cloned, some leaders realized that now might be a good time to get the ban lifted on some embryo research.

For in that 1993 tempest-in-a-teapot, a few sane voices had emphasized that producing twinned human embryos had several advantages for human beings. First, it could produce more embryos for infertile couples who, for various reasons, had difficulty producing enough for their cycles of IVF. Second, twinning such embryos could be useful in the diagnosis of genetic disease by the method of testing one cell of an 8-cell embryo. If many tests needed to be done, many cells may be needed. Obviously, if you have 15 copies of the same embryo, tests are easy and the chances of harming the resulting baby are minimized by taking only one cell from each embryo (and not a bunch of cells from an 8-celled embryo that will later be implanted). So the least-harmed babies are created by permitting genetic testing on twinned human embryos.

Meanwhile, to look briefly at the bigger picture, over the fifteen years between the birth of Louise Brown in 1978 and the election of Bill Clinton in 1992, a lot had changed. During that time, infertility medicine had become one of the fastest-growing sectors in American medicine. In June 1993, with a more sympathetic Clinton administration, Congress revoked the regulations requiring EAB approval of embryo research.

Not at all giddy with its new freedom, the NIH formed the Human Embryo Research Panel January of 1994 to create guidelines on whether such research should be permitted and, if so, what kind. This Panel had 4 bioethicists, 2 lawyers, 7 scientists, and 6 members of various other backgrounds.[9]

Asked to divide possible research with human embryos into categories of acceptable for federal funding, unacceptable, and "warranting further review," the Panel found itself forced to hold its meetings in public. As a result, it was targeted by anti-abortionists who saw a slippery-slope link between allowing research on embryos and aborting human fetuses. Members of the Panel received in the mail graphic pictures of decapitated, mature fetuses. During the hearings, passionate anti-abortionists created a hostile environment, seeking control and a platform.

Under these circumstances, the Panel did a great job. It concluded that federal funding of research with embryos would improve the success and safety of procedures to reduce infertility, and that the present situation, where such federal funding was prohibited, would lower the quality and ethical oversight of such research. More important, as Ian Wilmut showed how to reverse in a ewe, understanding the basic processes of cellular differentiation and development (and after Dolly, reversal) could help physicians understand how pediatric and adult cancers developed. The Panel even concluded that, parthenogenesis, where an egg is stimulated to begin dividing without fertilization, might offer insights into ovarian cancers.

The Panel did not accept the distinction between so-called "spare embryos" and "research embryos." The former, already fertilized and left over from a couple's attempts at assisted reproduction, were already fertilized and would not be of aid in studying fertility. Embryos available from couples attempting IVF have higher rates of genetic abnormalities, so it was necessary to have original embryos from young couples to be used as "research" embryos. The Panel rejected as too controversial the twinning of human embryos before implantation or cross-species fertilization. But as mentioned previously, the Panel's recommendations were not, as most members thought they would be, accepted because they got caught up in so-called "partial-birth" abortions.

Panel member Professor Alta Charo concluded that, at least in the case of human embryos, "logical arguments are only rationalizations for gut feelings or religious viewpoints." She decided then that, "I don't think we can make good suggestions unless we understand what is compelling for the public."[10] It is likely that her view influenced her later NBAC colleagues.

Embryo Research Today:
The Craziness of the Hughes Incident

Just how crazy things have become was well illustrated by an incident at the beginning of 1997, when geneticist Mark Hughes got caught in a nasty, schizophrenic change in the National Institutes of Health (NIH) policy about embryo research. The previous ban on federal funds (especially NIH funds) has stayed in effect as conceptionists held the NIH's feet to its zealous fire, but research on embryos to prevent genetic defects was possible with private funds, such as from Planned Parenthood.

Mark Hughes made *Science* magazine's list of scientific breakthroughs for 1992, coming in as a runner-up for his work in genetics on human embryos. He had pioneered a technique for taking DNA out of a single cell of a human embryo and testing that DNA for cystic fibrosis (CF). As previously mentioned, there is no evidence that taking one cell from an 8-cell embryo with undifferentiated cells does any damage to the embryo. The importance of this test is obvious to the 1 in 22 whites in this country who carry a CF gene.

Obviously, Hughes is a person at the forefront of scientific research helping humanity. He is a hero, not a villain. NIH thought so in 1992 when it enticed him to come to Washington, D.C. from his previous lab at Baylor's College of Medicine in Houston. One hitch in the plan was that, instead of working at the NIH lab in Bethesda, Hughes would work at Georgetown University, which had an NIH contract for a National Center for Human Genome Research. One wonders in retrospect about the wisdom of putting such a center at a Catholic university.

Hughes' goal at his new lab in Washington was to discover new genetic tests that can be done on human embryos. He could thereby enable couples with hereditary risks of genetic disease to use in vitro fertilization to test multiply-conceived embryos for diseases such as sickle-cell anemia, Down syndrome, and—with new discoveries of genetic diseases trickling in monthly from the Human Genome Project—perhaps one day treat such diseases.

Hughes was also appointed to the above 1994 panel that formulated public policy for embryo research. As mentioned, that panel had concluded that some modest research on embryos, such as that done by Hughes, was morally permissible. The panel had spent a great deal of time marshalling facts and evidence to support its modest conclusion. Certainly it expected its conclusion to be accepted. Certainly Georgetown University expected the panel's conclusion to be accepted when it hired Hughes, knowing that he did research on embryos. Certainly NIH expected the same.

But, as we know, the panel's recommendations were rejected. The ban on partial-birth abortions later passed the House in 1997 by a huge margin. This was an emotional issue to the folks back home, and seasoned politicians knew that by pressing a few of such buttons—ones about gay rights, abortion, and teenagers on welfare—you could throw a smokescreen over those contributions you took and what you did in return.

So it was a great alarm and surprise to the research community when *Chicago Tribune* reporter John Crewdson reported on January 9, 1997, that NIH had "terminated" its relation with Hughes because this researcher—"contrary to NIH policy, contrary to NIH instruction"—had been using NIH resources to test DNA from eight-cell human embryos. What else did NIH think Hughes had been going to do when it brought him to Washington?

In truth, the political heat had been turned up on NIH about embryonic research. NIH is a political animal and does not bet the agency on core principles; instead it follows utilitarian logic and is willing to sacrifice a few million here and there to keep its overall, multi-billion dollar budget. For example, all scientific and medical reviews indicated in the early 1980s that the Jarvik artificial hearts were mechanical stupidities that at best prolonged the dying in terminal cardiac patients and at worst tortured such patients in gruesome death rites in surgical intensive care units. (Jack Burcham left the OR with his chest only partially closed around the device because it was too large to fit; he died hours later, having been killed by the operation.[11]) NIH stopped funding artificial hearts in May 1988, but that was before Senators Hatch and Kennedy realized that such a cessation would mean the loss of $22.6 million to heart research centers in Utah and Massachusetts. Two other states, Texas and Ohio, were also affected, and when those states' congressmen were enlisted, an NIH official said (anonymously), "With all that Congressional pressure and the threat of legislation, we felt that the heart institute better eat a little crow rather than risk the future budgets of all the institutes."[12]

When fundamentalists added a clause to NIH's appropriations bill in 1996 barring federal funds for research on human embryos, NIH saw the writing on the wall. Previously, it had discreetly ignored the federal ban on embryo experimentation. More charitably, it had interpreted work such as Hughes' as not putting embryos at risk, and of course, there is no evidence that Hughes' work would put at risk an embryo implanted into a womb for gestation. But what about the embryos not implanted in Hughes' work? That was the flash-point.

NIH suddenly "remembered" that not every embryo in Hughes' research was implanted. The addendum to the NIH appropriations bill stated, "None of the funds made available in this act may be used for . . . research in which a human embryo or embryos are destroyed, discarded, or knowingly subjected to risk of injury or death greater than allowed for research on fetuses in utero."[13]

Now that these embryos had been remembered, American embryology would become stagnant while research in the field ground to a halt. For how do you do research in embryology if not on embryos? What's the point of billions of research-hours spent on animal embryos if it can never be applied to humans?

Hughes had tried to work out a compromise by putting part of his work across the street from his NIH lab at a private organization, Suburban Hospital in Bethesda, Md. According to *Science* magazine, "There, in separate facilities, he conducted his NIH-funded single-cell DNA studies and his embryology work, with the full knowledge of Georgetown and NIH."[14]

The absolutely incredible thing about all the brouhaha is that those who see themselves in the white hats may be the ones on the dark side. What was Hughes doing that was so terrible in his lab? Was he a mad Dr. Frankenstein, laboring away in the wee hours to create a mindless, but physically superior, Master Race, obedient to his sinister will? Not at all. He was helping couples, who used in vitro fertilization to create multiple embryos, to avoid familial genetic disease. Hughes would take DNA from one cell of each embryo to his private lab at Suburban Hospital, where tests would be run for diseases as spinal muscular atrophy, Tay-Sachs, or cystic fibrosis.

What do these Keepers of the Holy Embryos say to parents who now must have a child born with spinal muscular atrophy? "Yes, it's true. We could have tested your embryos and only implanted a healthy one, thereby not having this baby, who will live always in a wheelchair until he dies painfully as a teenager. But it's God's will." (At least, as We, the Appointed Ones, interpret His will to you and for you.)

Officially, Hughes got into trouble because some post-doctoral fellows worked sometimes in his private lab, and these post-docs were NIH-funded. One postdoc told some NIH official that he feared he was doing research that was banned. So Hughes said that, from then on, he would do such research himself.

Then came the refrigerator incident. A refrigerator assigned to Hughes' Georgetown lab had ended up in the lab at Suburban Hospital. For this egregious sin, the boom was lowered on Hughes. (How much does a refrigerator cost? Here's a man making well over $100,000 a year with millions in grants. If there's a problem, he can go to an appliance store and just *buy* another one). And why focus on the 'fridge? Because it was the "scene of the crime" where the embryos were stored? Officials at Georgetown's genome center gravely concluded that NIH resources had been "diverted" for (seemingly) clandestine embryo research in the other facility. Shortly thereafter, on October 21, 1996, Hughes and NIH agreed to end their relationship, and all his NIH funding ceased for any research on embryos.

Support from Georgetown was not forthcoming: "We take this matter very seriously and are continuing internal discussions about what to do," intoned a GU spokesperson. Harumph! What did they think he had been doing all along, especially when he helped Georgetown get grant money? The Hughes incident has had not just a chilling effect on embryonic research but a freezing effect: "The unfortunate fact is that . . . the brightest people, who really should be doing this sort of thing, aren't, because it's not legitimized science."[15] Not "legitimized" because of the lack of stable, long-term funding and because of the politics, not because there are not important discoveries waiting to be made.

The Importance of Wilmut's Discoveries for Understanding Humans

It was widely reported in the media that Wilmut needed 244 attempts to get one live lamb, but that number must be put into a realistic context. Of 244 nuclei transferred to eggs and put in lambs, only 34 were actually transferred into Scottish blackface ewes (the rest did not develop enough to do so).[16] After a week, he decided that 19 were healthy enough to proceed and discarded the other 15. From these 19, he got 5 identical Welsh mountain lambs. Two died within 10 days.

Of course, human embryos might not act the same. One of the things that may have enabled Wilmut to be successful was that sheep embryos begin to form specialized cells—what embryologists call "differentiation"—relatively late compared to other species. Embryos of mice differentiate quickly, whereas human embryos start later than mice but before lambs.

New Calls for Banning Research on Embryos

As mentioned above, the ancient fear is that society cannot draw lines between allowing research on one part of a continuum and proceeding to the end point. Fears of slippery slopes of justification abound about research on human embryos.

Leon Kass made such fears explicit in an article that appeared just before the announcement of the recommendations of the National Bioethics Advisory Commission. "And yet, as a matter of policy and prudence, any opponent of the manufacture of cloned humans must, I think, in the end oppose also the creating of cloned human embryos."[17]

Because he fears that allowing cloning of human embryos will inevitably lead to implantation of a human embryo originated by cloning, he wants to "test" physicians and scientists who favor lifting the ban on embryo research by making them endorse "an absolute and effective ban on all attempts to implant into a uterus a cloned human embryo (cloned from an adult) to produce a living child."

To the criticism that the techniques of human asexual reproduction are not that complicated and that someone in the world will eventually originate a living child by cloning, Kass would put the onus of proof on those who would permit the "horror" of such origination:

> Perhaps such a ban will prove ineffective; perhaps it will eventually be shown to have been a mistake. But it would at least place the burden of practical proof where it belongs: on the proponents of this horror. . . .[18]

But we know what happens if this line of reasoning is followed. We have to prove beyond a shadow of doubt to everyone that origination of humans will only have good results and that no parents anywhere will ever make a foolish choice about when and how to conceive. Unless we can prove this, we should not allow a human to be originated by NST, and until we can justify doing so, we should not allow human embryos to be twinned to study embryonic development or genetic disease. Obviously, such Olympian requirements are merely designed to block a process that some people find hard to accept.

What Is Lost in Research

At the end of Chapter 2, I discussed how gene-centric views dominate popular culture and that a more sophisticated view of genes sees them as variable functions that interact with the environment to cause an individual's phenotype. I also mentioned that even embryogenesis depends on not just the variable functions of the genes but the variability of the uterine environment.

How all this works in human embryogenesis is never going to be known, precisely, until we decide to do research on human embryos. Ian Wilmut's very achievement is testimony to how useless is a priori reasoning in this area. Wilmut overcame "known truths" and turned them into falsehoods. What is true or false about human embryogenesis—how much is the "in the genes" and how much is not— will never be known precisely until we do experiments with such human embryos. To say we can never do such experiments because such embryos are "tiny persons" or because it violates the "sanctity of life" is to say that humans are never meant to know truths about how embryos develop, how genes regulate such development or fail to do so in deleterious ways, and how it all goes together with the uterine environment to create a baby's phenotype.

Recommendations of NBAC

The National Bioethics Advisory Commission has recommended that research be permitted on human embryos, but not on any human embryo that might be gestated and brought to personhood. Will its recommendation be followed by NIH or, given the present climate about partial-birth abortions, will NIH allow such research to stagnate? The history of American research on the subject of research on embryos does not make one optimistic.

Conclusion

The history of American views on the moral status of the human embryo is not encouraging. There has been an unwillingness to face up to the fact that such embryos are not persons. The views of those who think the contrary are holding hostage important medical research. Whether it will be possible now to do such research in a limited way, as NBAC recommended, remains to be seen. It would certainly be a desirable outcome.

Notes

1. Sheryl Stolberg, "Reproductive Research Far Outpaces Public Policy," *Los Angeles Times*, 29 April 1997.
2. NBAC, *Cloning Human Beings: Report and Recommendations of the National Bioethics Advisory Commission*, Rockville, Md., June 1997, 42.
3. Joseph Fletcher, "Ethical Aspects of Genetic Controls," *New England Journal of Medicine* 285 (30 September 1971), 776–783. Reprinted in *Humanhood: Essays in Biomedical Ethics* (Buffalo, NY: Prometheus Books, 1979), 88.
4. Associated Press, "Pre-Pregnancy Screening Yields Baby Free of Fatal Genetic Defect," 22 January 1994.
5. A. Wilcox et al., "Incidence of Early Loss of Pregnancy," *New England Journal of Medicine* 319, no. 4 (28 July 1988), 189–194. See also J. Grudzinskas and A. Nysenbaum, "Failure of Human Pregnancy after Implantation," *Annals of New York Acad-*

emy of Sciences 442 (1985), 39–44; J. Muller et al., "Fetal Loss after Implantation," *Lancet* 2 (1980), 554–556.

6. Joseph Fletcher, "The Cognitive Criterion of Humanhood," *Hastings Center Report* 4 (December 1975), pp. 4–7.

7. Mary Ann Warren, "On the Moral and Legal Status of the Fetus," *The Monist* 57 (1973), 43–61.

8. Paul Badham, "Christian Belief and the Ethics of In Vitro Fertilization," *Bioethics News* 6, no. 2 (January 1987), 10.31.

9. Ronald M. Green, "At the Vortex of Controversy: Developing Guidelines for Human Embryo Research," *Kennedy Institute of Ethics Journal* 4, no. 4 (December 1994), 345–356.

10. "Little Known Panel . . ." , *New York Times*, 19 March 1997.

11. Gregory Pence, "Artificial Hearts: Barney Clark," *Classic Cases in Medical Ethics*, 2nd ed. (New York: McGraw-Hill, 1995), 282.

12. Phillip Boffey, "Federal Agency, in Shift, to Back Artificial Heart," *New York Times* 3 July 1988, A1.

13. HHS Appropriation, 1996, quoted in *Science* 275, no. 24 (24 January 1997), 472.

14. *Science* 275, no. 24 (24 January 1997), 472.

15. *Science* 275, no. 24 (24 January 1997), 472.

16. T. Adler, "Bidding Bye-Bye to the Black Sheep?" *Science News*, 9 March 1997.

17. Leon Kass, "The Wisdom of Repugnance," *The New Republic*, 2 June 1997, 26.

18. Leon Kass, "The Wisdom . . . ," 26.

Arguments for Allowing Human Asexual Reproduction

It is depressing, not comforting, to realize that most people are accidents. Their conception was at best unintended, at worst unwanted. There are those who are so bemused and befuddled by a fatalist mystique about nature with a capital N (or "God's will") that they want us to accept passively whatever comes along. Talk of "not tinkering" and "not playing God" and snide remarks about "artificial" and "technological" policies is a vote against both humanness and humaneness.

Joseph Fletcher, 1974[1]

· · · · ·

In this chapter, I examine positive arguments for originating a human by asexual reproduction. From this point onward, and in order to neutralize some of the emotions surrounding this topic, I shall often speak of human nuclear somatic transfer (NST) instead of human asexual reproduction. Also, and for reasons of simplicity of argument, I shall concentrate on the simple, clear case of parents who desire just one child. The following hypothetical case is another example of where such desires might be realistic.

Case #2—John and Elsie Kennedy

John and Elsie Kennedy had been trying to have a child for several years before they went to the infertility clinic. There it was determined that John had no sperm. John had once worked at a power plant run by nuclear energy and it is thought that exposure to radiation has rendered him sterile.

This has caused a crisis in the Kennedy marriage. Although Elsie was only mildly in favor of children, John passionately wanted them. This causes him to be very depressed. They consider artificial insemination by (anonymous) donor (AID), until Elsie says, "Why not clone John's genes into my egg? It's better than some unknown guy's! And John will have a great connection to the child, as he always wanted. We know he won't be John, but it's certainly better than surrogate donation."

Personal Liberty and the Right to Self-Reproduce

I do not tell you how many children you should have or whether you should have children at all. Neither does the government. If the federal, state, or county government attempted to do so—if it attempted to say you should get pregnant, or should have at least five children (as the Nazis attempted), or should get sterilized because you were the wrong kind of person—you would rightly object to this as an offensive, grave violation of your personal liberty.

The essence of democracy is that government is not a reproductive dictatorship. People in it have real freedom in their personal lives about having babies. The "right to privacy" invoked in legal reasoning is a misnomer, for the issue here is not usually private, although it is certainly personal. The right to privacy would be better called "the right to personal decisions." Certainly, decisions about reproduction are in this sphere.

This is the essence of John Stuart Mill's harm principle: the freedom to be left alone by others and by the government. James Hughes, a sociologist at the University of Connecticut at Hartford, puts it this way: "We allow people to make reproductive choices, and this [originating a human by cloning] is a further extension of that. If a parent wants to clone a dead child as an act of love, it may be neurotic, but is it any less neurotic than wanting another child to replace a dead child?"[2]

Medical ethicists at the University of Michigan and Michigan State University received a grant from the ethics branch of the Human Genome Project to conduct a series of meetings across Michigan to discuss issues raised by genetics.[3] They focused on four questions:

1. How expansive should the domain of reproductive liberty be with respect to genetic screening?
2. How expansive should the domain of reproductive privacy be with respect to genetic screening?
3. How much should society subsidize the costs of genetic screening for those who cannot afford it?
4. How should the concept of genetic responsibility to future generations be implemented in public policy?

Among the consensual conclusions of these meetings was that the genetic privacy of people should be respected, that people should not be forced to know if they are at risk for a genetic disease, that genetic privacy of children should be respected if there is no treatment available, and that engaged couples should not be morally obligated to undergo genetic testing to "define their genetic endowment and to identify any genetic risks" to their children.

Another very interesting conclusion, endorsed by 60–70% of people, was the statement that, "Parents should be morally free to pursue whatever alternative reproductive technologies are available to avoid the birth of a child with a serious genetic disorder." Although participants were not specifically considering human NST, this principle is a broad, attractive one that could apply to creating a child by NST.

Legally, it is not at all clear that the right to reproduce by NST does not already fall under an American's right to reproductive liberty, a view emphasized by law professor John Robertson.[4] The U.S. Supreme Court has made numerous decisions clarifying how the right to privacy, if it means anything, means the right to control decisions about personal reproduction. In particular, the Court's 1972 decision in *Eisenstadt v. Baird* said,

> If the right of privacy means anything, it is the right of the individual, married or single, to be free from unwarranted government intrusion into matters so fundamentally affecting a person as the decision whether to bear or beget a child.[5]

This is very strong language. In specifically emphasizing the rights of the "individual" to "beget," such language creates a direct connection with asexual reproduction. Other such legal decisions have specifically mentioned that the right to privacy not only includes the right to stop pregnancies but also the right to undergo medical procedures to create pregnancies.[6]

Benefit to Children—Improving Genetic Inheritance

The strongest direct arguments for originating a child by NST is that his parents might give him or her a wonderful genetic legacy. The idea of doing so has been so corrupted by associations with half-baked proposals from the past that the barriers to considering it pop up almost immediately. Yes, there might be mistakes in trying to give children a better genetic start in life (but there are mistakes in choosing schools, in trying to plan conception of children, in estimating one's capacity to be a good parent, and such mistakes don't justify a policy that bans children). Yes, there is a danger of society forcing parents to strive in this direction (but such a danger is nowhere on the landscape or even on the horizon now, so why act as if such dangers lurk overhead?). But before we get to all the "buts," can't we keep an open mind about how children might not be harmed but might—very likely—be benefitted by having their parents deliberately seek a better genetic heritage for them?

Future people can just as equally be affected by whether we (a) do things that harm them, or (b) omit to do things that could have benefitted them. We can harm future people by encouraging people with lethal genetic diseases to have as many children as possible, and then encouraging those people to have more children, e.g, by encouraging people in institutions for the severely retarded to have children. To appreciate just how much harm can be done, consider one documented case in Venezuela, where a sailor with Huntington's disease jumped ship in the early 1800s and had children with a local woman. By 1981, this sailor's descendants numbered over 3,000 (of which over a thousand were then at risk for Huntington's).[7]

Consider the following case, which is unfortunately like hundreds of thousands of similarly true cases that occur each year on this planet.

Case #3—Richard Dunaway

When Richard Dunaway, a white male aged 58 years old, drove his tractor-trailer into Nu-Way Trucking in Alexander City, Alabama, he got out and the right side of his body felt so weak that he could hardly stand on his leg.[8] Inside the trucking terminal, he couldn't remember his home telephone number. In a second, his whole life had changed, and in many ways, his real life was over. When someone drove him home, he couldn't remember the name of his wife, Sandra, and he looked at his 5-year-old granddaughter, Skye, and asked, "What's your name?"

The cause of Mr. Dunaway's problem, diagnosed at the University of Alabama at Birmingham, was glioblastoma, a deadly malignant cancer of the brain. Richard was given 12 to 18 months to live. Cancer runs in Richard's family: his mother had colon cancer and his father had already died from the same. There is very likely a genetic basis for so much cancer in one family, in the same way that ovarian and breast cancer runs in families.

The terminal diagnosis devastated Sandra Dunaway, who had been married to Richard for 37 years: "He's the only breadwinner in our family. All of a sudden, it's like everything has just come all to pieces. I'm hurting, but it's inside." Mrs. Dunaway worried about their youngest daughter, who was

19 and retarded, and about how they would get by without Richard's paycheck.

Although we do not know how much early death is caused by heritable diseases, a lot is. Until we try to prevent such deaths, we will never know how long humans can live.

For the other two million Americans who die each year, the statistics aren't much better: 750,000 from heart diseases, 500,000 from cancers, and 150,000 from strokes. Over 70% of deaths may be from preventable, genetic causes. Seen in this light, originating humans by NST from healthy adults doesn't seem quite so crazy.

One might ask, why is part of the solution human NST rather than genetic therapy? The answer is that human NST may become a tool to prevent common genetic diseases. Somatic genetic therapy in humans has been progressing very slowly and may take many decades to perfect. In the meantime, we need to be able to use every tool possible to prevent genetic disease. Once studies in several species of mammals prove it safe, human NST may become a good tool for those fearing familial, genetic disease.

Risks of Common Genetic Diseases[9]

Condition	Estimated Incidence
Cystic fibrosis	1 in 25–32 (carriers of CF gene)
RH incompatibility	1–2 in 100
Diabetes	1 in 80
Sickle cell anemia	1 in 625 African-Americans
Polycystic kidney disease	1 in 1,000
Down syndrome	1 in 1,050
Beta-thalessemia	1 in 2,500
Tay-Sachs disease	1 in 3,000 Askensasic Jews
Huntington's disease	1 in 5,000 (heterozygote frequency)
Duchenne muscular dystrophy	1 in 7,000
Hemophilia	1 in 10,000 males
Phenylketonuria (PKU baby)	1 in 12,000 Caucasians
**Breast Cancer	1 in 10 females (discovered or suspected genes; probable genetic co-factors)
**Prostate Cancer	1 in 10 males (discovered or suspected genes; probable genetic co-factors)

For those who think that using NST to create a child would be to manufacture humans or to tamper with nature, consider how each new "Richard Dunaway" feels when he is diagnosed with cancer and thinks that he could have been born with the probability of living 20–40 more years. The real Richard Dunaway's birth and his lethal inheritance was due to the arbitrary whims of fate; that need not be true for those of the next generation.

Consider a case that might be true one day but which is now fiction.

Case #4—Robert Atworthy

Robert Atworthy's parents, Mimi and Harrison Atworthy, wanted to give Robert the finest genes they could. They selected Mimi's grandfather, Herman, who was 90 and who never had had cancer, heart disease, or cerebrovascular diseases. They wanted to use a senior relative to keep the genetic ancestor in their family. Herman died when Robert was eight, and Robert dreams sometimes of talking to Herman to "thank him for my genes."

Robert is presently 95 years old, thinks clearly, walks daily. Because of his longevity, he was able to collect three different pensions: one from the Armed Forces, another from his job at a large company, and after he took early retirement from that job, from a Keogh plan he set when he founded his own small business making bagels.

Robert has been married to the same woman for 60 years and knows all his 17 grandchildren and 18 great-children. Although most of his children did not choose to replicate a pre-existing genotype, some did, and three of his grandchildren and 9 of his great-children have Robert's genes. It is predicted that a human baby born with his genes now has a life-expectancy of 120 years.

Says Robert, "I thank my parents, Mimi and Harrison, for what they did. At the time, they were real pioneers. I figure I've gotten 35 extra years of good life because of what they did. I didn't really start to slow down until I was 110. I'm glad that some of my grandchildren will have the same base, and with better medicine and drugs, some of them might make it to age 150."

Although such a case seems contrived now, there is no reason that it could not come about in a benign, non-coercive way through the voluntary decisions of ordinary couples.

The possibility of reducing risk of genetic disease for a child by selective choice of ancestor-genes is the strongest direct argument in favor of originating a child by NST. It directly counters a major objection that such origination may harm the child.

Failure to appreciate this argument is often due to a failure to understand the present great toll from genetic disease in early death and a similar failure to imagine just how great the benefit could be to future people of selective asexual reproduction.

There are people in medical genetics and medicine with much stronger views than the one expressed here, people who have all their professional lives seen the terrible results of genetic disease. For example, respected genetic researcher Marjery Shaw once suggested that deliberately giving birth to a child with the gene for Huntington's disease should be a criminal offense.[10]

Benefit to Children—Genetic Therapy/ Correcting Genetic Disease

In his testimony to Congress, National Institutes of Health Director Harold Varmus tried to counter the effects of George Annas' negative testimony. Varmus said, "We should contemplate whether there are rare situations where we should consider human NST."[11] He mentioned cases of genetic disease and infertility.

As with in vitro fertilization, the Gloomy Gus mentality delayed attempts at somatic genetic therapy for nearly twenty years. Finally, after much hand-wringing, endless debate, and many years of genetic testing on animals, somatic gene therapy was first attempted on September 14, 1990 at the National Institutes of Health (NIH). The subject was a 4-year-old girl with adenosine adaminase deficiency (ADA), a fatal disease resulting from lack of an enzyme, and scientists attempted to insert the genes causing this enzyme into her. There were no controls. Three years later, the girl was still alive (without the new gene, she would have been dead), and attending school. However, the results were not as promising as had been hoped.

Before NST came along, the next logical step was germ-line or gametic therapy, which can be done in two forms: by altering the genomes of embryos or by altering the germ cells of a couple, such that altered germ cells then produced genetically-altered gametes. The latter is much riskier than embryonic modification because it is impossible to know if all of the targeted germs cells have been altered as intended. It is here that the "Jurassic Park" objection is strongest: we simply would not know what future kids would be like if we started making genetic changes in their parents' germ cells, and any mishaps could cause real human suffering.

When we turn to modification of embyronic cells to avoid genetic disease, it seems to me that there is no reasonable objection. One day we may be able to prevent—for couples rich and savvy enough to use in vitro fertilization—cystic fibrosis, blindness, deafness, dwarfism, spina bifida, and Down syndrome. Is there anyone who seriously thinks either that children are better off with these conditions or that parents should not have the right to alter the genes of their embryos to eradicate such conditions?

Enter NST. Many experts worry that germ-line genetic therapy is too risky, but NST may be an attractive alternative. It is important not to prejudge this issue and to consider all the alternatives. For example, right now a very high number of people consider abortions justified when amniocentesis or sonograms reveal a genetic disease. What people forget is how painful the choice to abort such a fetus can be. The test usually does not come back until some time in the second trimester and by that time, the couple has been thinking about the baby for four to six months. If it is a first-time pregnancy, or a pregnancy in an older couple that was difficult to achieve, it may be very difficult for them to choose to abort. As bioethicist John Fletcher has argued in many forums, such a choice traumatizes couples because they are not choosing against being pregnant but against a particular genetic condition.[12]

Originating a child by NST would be especially valuable in cases of autosomal dominant genetic diseases having a 50-50 risk of inheritance, as well as certain lethal sex-linked diseases, and for which there is presently no presymptomatic test. A child originated from the non-affected parent would at least be free of the feared disease.

Moreover, even if somatic therapy was effective, the germ cells of those affected with lethal genetic disease would still contain the genes for the disease. People such as Nancy Wexler, a famous researcher who helped discover the gene for Huntington's disease and who herself is at risk for developing this disease, chose not to have children. How much better to have children, originated by NST, and know they are free from Huntington's.

One might object that Nancy Wexler could also have children by using pre-implantation diagnosis (PID) of embryos created by IVF, but the procedures of IVF, done in addition to the procedures of PID, are very expensive and quite uncommon. If the bans on both embryonic experimentation and human NST were lifted, inexpensive, reliable methods might be developed that were alternatives to IVF/PID.

Aiding Infertile Couples

NBAC did not specifically recommend a federal law forbidding creating embryos by NST with the intention of not bringing them to birth. Because such creation is forbidden with use of federal funds, and because the 1994 Embryo Panel *Report* already existed, NBAC preferred to finesse issues dealing with embryos. But there are scenarios where originating embryos by NST could benefit infertile couples. Moreover, if such scenarios occurred in private clinics not using any federal funds, absent a new federal law to the contrary, they could be legally carried out right now.

A woman ordinarily ovulates once a month where one egg gets primed for possible conception. Infertility doctors must surgically remove such eggs by a lighted, tubal instrument called a laparoscope. In order to minimize the number of times such surgery is done, the physicians give drugs to such women to stimulate superovulation, so the physicians can retrieve many eggs in one operation. Such drugs are thought by some to cause increased risks of cancer for such women in their later years. If physicians could remove one egg, produce an embryo, and clone that embryo to create multiple copies, they could avoid having to give women such drugs (and such drugs are often given to them month after month).

In vitro fertilization is a very expensive process, costing about $8,000 per attempt in 1997, and not covered by most insurance policies. In part it costs so much because of the use of such drugs and because of repeated surgery to retrieve eggs. By using twinned embryos, costs might also be considerably reduced and the whole process could be made much easier.[13] This is not NST from an adult, differentiated cell, but it is presently forbidden and should not be.

IVF seems to have hit a wall for older women in that success rates with the eggs of these women are not great. The real movement now in fertility medicine is using the eggs of younger women, which has greatly increased rates of success. But even that change has limitations. Early in 1994, newspapers in England erupted over a report that eggs could be retrieved from aborted female fetuses, implanted in infertile women, and artificially ripened for conception. This extreme measure was apparently being considered because in England donated eggs from younger women are scarce: infertile women who need donated eggs now face a 3-year wait.[14] The scarcity of eggs in England underscores the need for new ways to help those who want to have kids and who cannot: we keep getting back to the fact that, because of a minority's religious view that personhood begins at conception, infertile people who want children are being denied new ways of trying.

As said, we know that 40% of embryos in sexual reproduction fail to implant. One possibility here is to twin embryos of several different genomes, implant one of each, and see which successfully implants. Those that are successful could be identified and set aside for "adoption" by other couples, on the assumption that having already successfully come to term once, they are more likely to do so again. In addition, by waiting a few months, it might be determined that the resulting children were completely healthy. This is a form of asexual reproduction now probably banned, but to what end?

In the past, conservative critics of in vitro fertilization such as Leon Kass have argued that "it might be better for such infertile couples to wait" until the safety of new techniques are conclusively proven. A similar argument has been advanced by the National Bioethics Advisory Commission. But to such arguments, one infertility physician replies, "None of these people [these critics] have sat down and talked to my patients. None of them have seen the misery my patients are living through."[15]

Valuing the Genetic Connection

In this section, I set up an argument that is completed in the next. In discussions about having children and raising them, many people assume that having a genetic connection between parents and children is valuable and that society should promote it.[16]

Such a person might argue that, "the genetic bond creates the strongest possible bond between parents and children. Parents take their responsibilities more seriously when there is a genetic connection. When fatherhood is questioned in a pregnancy, the courts and men themselves acknowledge paternity when tests establish a genetic connection."

Many people worry today about the difficulties of being a good parent. One in two marriages ends in divorce. More mothers than ever must work in order to get by, resulting in less time at home with children. With latchkey children, children who have been mostly raised in day-care centers, and the stresses of blended families, children are more at risk than ever. Most of the thirty-plus million Americans who do not have good medical insurance are children.

In such dire straits, society should not be too eager to throw off the traditional supports of parenting. One such support is the biological-genetic basis of the relationship. If we downplay that relationship, if we publicly say it doesn't matter, and if legislatures act similarly, we risk destroying the foundations of the family.

Because of the importance of the genetic connection, judges in the Baby Richard and Baby Jessica cases have returned adopted children to a biological parent. Although these judgments have caused a great outcry that such actions are not the best thing for the particular child, the critics may have adopted a myopic viewpoint.

It is one thing to ask what is best for this particular child and to ask who are the best parents for it. It is quite another to ask about what is the best law governing all such cases. If we adopt a law favoring biological parenting, then of course there will be hard cases where heartbreak occurs by enforcing the law. On the other hand, if we allow too many exceptions to the law, we don't really have a law anymore.

The genetic relationship also matters to those in the extended family who are not the parents. If both parents should be killed in an accident, the grandparents, uncles, and aunts of the children feel obligated to raise the orphans in a way that they do not feel for orphans of strangers. Whether an uncle feels this way because he has come to know his nephews and nieces better than other children, because society has socialized into him a feeling of responsibility for them, or because he really feels some obligation because of a genetic connection, is irrelevant to a public policy that wants newly-orphaned children to be adopted.

The importance of biological connections is also seen in how women are more connected to babies than men because of child-bearing. Men are not as connected to babies as females during gestation, and also not as connected after birth when mothers are able to breast-feed. It is only after these experiences are over that men and women bond equally to children, but at this point, women are far ahead of men. In cases of divorce, is this why far more women than men want custody of children and want to rear them?

Critics of a genetic bond allege that couples who want their own kids are genetic narcissists, and maybe even selfish, sexist, or racist. (Having one's "own" kids always seems to mean having kids who are of the same racial and ethnic background.) It is immoral and irrational to view children as one's property, as being expressions of one's identity, and as being substitutable for a career. Children should be seen as Kant saw people, as "ends in themselves," as moral agents, and as autonomous beings worthy of respect simply because they are people. Kids do not exist for parents and should not be created for them. Parents exist for kids, and any public policy that says the opposite is bad for kids.

Critics of the genetic bond also claim that couples who worship the genetic connection are ill-informed because there is not the connection they think. At best, a child only has half of one's genes, and the genes he inherits from John Doe may not be those that distinguish John's body, personality, talents, or intelligence from others. If we look at the roughly 100,000 genes that exist in humans, and if we compare the genes of two biological parents and their child with the same parents and an adopted child, the degree of difference is not as great as many might suppose. The alleles (combinations of genes) of both children will be very different from the alleles of either particular parent. We are talking differences of degree here, not of kind.

Defenders of the importance of the genetic connection believe that biology is hard to ignore and impossible to dissolve. If the genetic connection isn't important, why do adopted children characteristically go on quests to identify and meet their genetic parents? Is it merely a fad or something deeper? Why do children of uncertain origins seek to discover their real biological parents? Witness also the efforts people go through to attend reunions of extended families and to trace their genealogical roots. If genetic ties to long-departed ancestors weren't important, why would people do this?

Richard Dawkins, author of *The Selfish Gene*, argues that there is a good reason in evolution why humanity has arrived in the modern world with families who cherish genetic connections.[17] In times of scarcity of food or other resources necessary for survival and reproduction, families survived who took care of their own first. Those with genetic connections got shelter from storm; genetic orphans perished.

That, at least, is the view of sociobiology and its contribution to ethical theory. Another ethical theory, emotivism, argues that there is no objective truth in ethics and that rationality places a very small part in moral life. Perhaps emotivism best explains here why couples feel so connected to genetically-related children and why such beliefs are so immune from rational criticism. Rational or not, strong feelings about one's kin seem to create strong feelings of bonding between people. Perhaps there really is such a thing as bonding between mothers and children. Adopted children may be harder to bond with.

Some people feel that a genetic connection to children is a form of immortality. True, it does not prevent their death, but something of them continues into the future, albeit only half of their genes distributed in a random fashion. Nevertheless, something is better than nothing. Leaving children in the world gives such people a feeling of continuation of life as well as mastery over otherwise intolerable feelings of despair at the prospect of the end of their own lives.

Generalizing the Genetic Connection

Many of the arguments above for a genetic connection in sexual reproduction also support a genetic connection through NST to an ancestor. If it is good for a parent to have a genetic connection to a child, it is also good for a parent to have a very strong genetic connection to a child.

As argued, a genetic bond makes parents take their responsibilities more seriously, allows men to acknowledge paternity (and, if they don't, they are forced by the courts to do so), makes parents care more about what kind of future they leave to their children, and strengthens fragile family ties in a time when parenting is under stress.

One point in our former discussion is of special interest. We noted that women experience a bond with children during gestation and breast-feeding, but men do not. Many people have speculated that such experiences explain why in cases of divorce, women want custody of children more than men and take their responsibilities to children more seriously.

This raises the question of whether originating a child from the father's genes might actually strengthen a marriage because it would allow both parents to have a strong biological connection from the beginning. For the woman, there is the biological connection of becoming pregnant, carrying a growing fetus for nine months, giving birth, and breast-feeding. For the husband, it would be the extraordinary feeling of seeing his own genes originate again in a new context and with himself as father/sole ancestor.

In his column in *Nature* about the implications of Dolly, geneticist Axel Cahn predicted that public opinion might "tend to legitimize the resort to cloning techniques in cases, where, for example, the male partner in a couple is unable to produce gametes." He stressed that although the woman did not contribute any genes,

Nevertheless, for a woman the act of carrying a foetus can be as important as being its biological mother. The extraordinary power of such "maternal appropriation" of the embryo can be seen from the strong demand for pregnancies in post-menopausal women, and for embryos and oocyte donation to circumvent female sterility. Moreover, if cloning techniques were ever to be used [in this way] the mother would be contributing something—the mitochondrial genome.[18]

So both husband and wife might feel a strong genetic bond to a child created asexually.

In a discussion of the genetic connection between adults who reproduce sexually, Judith Andre argued that one value of emphasizing the genetic connection between parents and children is that it gives value to the public policy that says, "It is very important that every child have adults who are bound beyond question to that child's welfare."[19] In the case of the male ancestor, we have the opportunity to strengthen the male "bond" to the newborn child and to bind that male "beyond question" to that child's welfare.

Before we consider some objections to this idea, let us at least acknowledge that modern society has a problem getting fathers to feel connected to their biological children. Too many men regard marriage as a package deal, such that when their marriage to the wife ends, their obligations also end to the children they produced with that wife.

As mentioned before, perhaps one reason for such feelings by men is that the woman has gestated, birthed, breast-fed and raised the children (perhaps while they were at work), and the fathers often have had little relation to small children, especially if the father works outside the home and the wife stays in the home. In our new scenario, having a child who looked like you and had your genes would not be so easy to walk away from. Indeed, it would take a male who was almost non-human to do so. So a genetic connection between a father and a son originated by NST would likely produce a very strong genetic connection, and hence, a stronger-than-normal social bond between father and son.

This relationship certainly takes care of one previous objection to the importance of the genetic connection between parents. That objection was that given all the alleles and genes in the human genome, the difference between genetic kids produced sexually and adopted kids is not so great and only a matter of degree. That degree with NST rises to nearly 99.9%, so it is no longer possible to say that the gene-originating parent does not in fact have a strong genetic connection.

In sum, whatever arguments are typically given for the value of genetic connections between parents and children—and more may be possible than I have given here—are also arguments for the value of allowing a strong genetic connection between one parent and a child. Indirect benefits may also accrue from this arrangement to the second parent if she is female and gestates the child.

Rawls' Argument

Justice, according to Harvard philosopher John Rawls, applies not to acts between individuals but most fundamentally to the basic structure of society. Rawls argues famously in his *A Theory of Justice* that the principles of justice that apply to the basic structure would be chosen in a hypothetical social contract (he calls this "the original position") where parties choose under a "veil of ignorance" about their position in society when the veil rises.

People have emphasized various aspects of Rawls' theory, but few have emphasized the following passage:

> I have assumed so far that the distribution of natural assets is a fact of nature and that no attempt is made to change it, or even to take it into account. But to some extent this distribution is bound to be affected by the social system. . . . it is also in the interest of each to have greater natural assets. This enables him to pursue a preferred plan of life. In the original position, then, the parties want to insure for their descendants the best genetic endowment (assuming their own to be fixed). The pursuit of reasonable policies in this regard is something that earlier generations owe to later ones, this being a question that arises between generations. Thus over time a society is to take steps to preserve the general level of natural abilities and to prevent the diffusion of serious defects. These measures are to be guided by principles that the parties would be willing to consent to for the sake of their successors. I mention this speculative and difficult matter to indicate once again the manner in which the difference principle is likely to transform problems of social justice. We might conjecture that in the long run, if there is an upper bound on ability, we would eventually reach a society with the greatest equal liberty the members of which enjoy the greatest equal talent.[20]

To the argument that we should not attempt to improve the human race, Rawls provides a framework for a cogent reply: if we were in the social contract—taking the long view of millions of people over many generations—and we did not know which generation we would inhabit when the veil lifted, we would choose to make the later generations as genetically talented as possible, compatible with the equal liberty of each to procreate in preceding generations.

It cannot be stressed too much that, on Rawlsian principles, state coercion has no place in improving the genetic heritage of the human race. For Rawls and most members of democratic societies, the first principle of civilized life is protection of our basic civil liberties. Any attempt to impose a procreative program on us violates such liberties. Equally, when the state says we can not reproduce in certain ways, it also violates our liberties.

Under the veil of ignorance, it is in our interest to allow our parents to create each of us with as much natural talent as possible, with the best genes, and with the best chance at a long, healthy life. It is important to stress that, if I understand Rawls correctly, then under his famous theory of justice, people are not just permitted to improve the genes of future children, they are *obligated* to do so. It is wrong to choose lives for future people that make them much worse off than they otherwise could have lived.

Children for Gay Men and Lesbians

Genetic connections to children have long been denied to gay men; lesbians who wish to reproduce must find an accommodating sperm donor.

The author is a friend of one lesbian couple who approached a gay male to be the male donor of their child, and he agreed. Moreover, as they wished, he is part of the child's life. But "Mama 1" and "Mama 2" would have chosen to originate a child by NST if it were possible, with Mama 1 supplying the genes and Mama 2 supplying the egg and carrying the fetus to term. Although some people will be shocked at this arrangement, it may be more common than many people realize.

Of course, this couple realizes that their child will almost certainly be heterosexual. Even if there is a genetic basis of homosexuality (the so-called "gay gene"), it accounts for only 2% of the population and their child would be unlikely to inherit it. However, if there is a gay gene and if both members of this lesbian couple have it, then a child created by NST from either of them might almost have it.[21]

Originating a child by NST in such a lesbian couple would not, of course, eliminate a genetic connection to a male. The child would have a genetic connection to the father of the woman whose genotype was used, and the child might one day seek out a relationship with his or her male ancestor.

Most studies to date have found, contrary to public prejudice, that children of gay and lesbian couples have no more problems than children of heterosexual couples. For example, a study by psychologist Charlotte Patterson of the University of Virginia in 1992 found that children of gay parents are almost always heterosexual.[22] Indeed, given the fact that about a third of the nation's kids live in households with only one parent, a two-parent lesbian home might be a benefit for many kids. Long-term commitment to a relationship and to childrearing are undoubtedly more relevant to being good parents than sexual orientation.

Of course, some will propose that NST should be available only to married couples. This means restricting NST to heterosexuals, since homosexuals cannot marry. However, that would be the only kind of medical treatment for which it is necessary to be married. No restrictions are now placed on unmarried, fertile, heterosexual couples, who may use all forms of medical help to conceive.

Perhaps this is exactly the kind of scenario that worries critics of human NST. But absent a good argument as to why being a lesbian is a bad thing, no reason exists to deny lesbians the same benefits of genetic parenthood that other humans so easily obtain.

The NBAC *Report* was discussed when it was issued in late June of 1997 at a meeting of the American Association for the Advancement of Science in Washington, D.C., where a panel on scientific freedom and human NST featured theologians who condemned human NST. The only dissent came from scientist Franklin E. Kameny, Ph.D., and Randolfe Wicker, director of a gay-rights organization.[23] These two objected to the negative, wishy-washy dialogue about originating humans by NST.

"Those of us who celebrate the advent of human NST technology are largely excluded from public debates on this issue," they complained. Wicker went on to challenge "the mindset which so limits this debate" and characterized the panel of theologians as "canaries all singing the same song to each other." Dr. Kameny criticized "the planners of this conference [because] on the two panels I've attended today, not one was a dedicated, ardent pro-human cloning advocate. That is unconscionable."

The two speakers ended by asking the panelists two questions. First, isn't access to NST a reproductive right, especially for infertile or same-sex couples unable to reproduce by other means? Second, given that many people think that human NST will in fact occur soon by someone in some country, isn't making laws against NST just wishful thinking?

Conclusion

I have argued that it is permissible to allow a couple to originate a child by NST. Allowing couples to have children by NST should be part of each person's general procreative liberty. A child may also be benefitted by being originated by NST and being free of many common genetic diseases.

To the extent that the genetic connection is valuable from sexual reproduction, it is also valuable between one parent and one NST child. In some marriages, originating a child by NST from the father may strengthen the marriage and allow both parents to feel a strong biological connection to the child.

If Rawls is right, we are obligated to improve the genetic health of future generations and nuclear somatic transfer is one tool that humanity has to do so. We should not ban such a tool without good reasons. Finally, origination by NST allows gay men and lesbians to participate in human reproduction, a good thing.

Notes

1. Joseph Fletcher, *Ethics of Genetic Control: Ending Reproductive Roulette* (New York: Doubleday Anchor, 1974); reprinted by Prometheus (Buffalo, N.Y., 1984), 36.
2. Shankar Vedantam, "Human Cloning: Some See Pluses," *Denver Post*, 13 March 1997.
3. Leonard Fleck, "Genetics and Ethics: A Community Dialogue," *Medical Humanities Report*, Center for Ethics and Humanities in the Life Sciences, Michigan State University. Spring 1997, 1–3.
4. John Robertson, "A Ban on Cloning and Cloning Research Is Unjustified," Testimony Presented to the National Bioethics Advisory Commission, 14 March 1997.
5. United States Supreme Court, *Eisenstadt v. Baird* 405 U.S. 438, 453 (1972).
6. *Lifchez v. Hartigan*, 735 F. Supp. . . . 1361 (N.D. Ill.), affirmed without opinion, sub nom.
7. See "Presymptomatic Testing for Genetic Disease: Nancy Wexler," in Gregory E. Pence, *Classic Cases in Medical Ethics: Accounts of the Cases that Shaped Medical Ethics*, 2nd ed. (New York: McGraw-Hill, 1950), 396.
8. Jeff Hansen, "New Site Man Thrust into Fight for His Life: Joins UAB Experiment," *Birmingham News*, 7 April 1997, D1, 5.
9. Incidence of these common genetic diseases is taken from the OMIM—Online Mendelian Inheritance in Man, web address: http://www3.ncbi.nlm.gov/omin/.
10. Margery Shaw, "Conditional Prospective Rights of the Fetus," *Journal of Legal Medicine* 63 (1984), 99.
11. Shankar Vedantam, "Human Cloning . . . ."
12. John Fletcher, Medical College of Wisconsin Medical Ethics listserv discussion, 24 March 1997. Quoted by permission of John Fletcher.
13. Gina Kolata, "For Some Infertility Experts, Human Cloning Is a Dream," *New York Times*, 7 June 1997.
14. E. Robinson, "Idea of Using Eggs from Fetuses Raises Furor," *Washington Post*, 17 January 1994.
15. Gina Kolata, "For Some . . . ."
16. I am indebted greatly in this section to a general discussion on the medical ethics listserv originating from the U. of Wisconsin on genetic connections to children that occurred between 30 August 1995 and 14 September 1995.
17. Richard Dawkins, *The Selfish Gene* (New York: Oxford University Press, 1976; 2nd ed., 1989).
18. Axel Kahn, "Clone Animals . . ."
19. Judith Andre, Medical College of Wisconsin Medical Ethics listserv discussion, 6 September 1995. Quoted by permission of Judith Andre.

20. John Rawls, *A Theory of Justice* (Cambridge, Mass.: Harvard University Press, 1971), 108.

21. Things of course will be more complex in reality. Perhaps 1% of the population might have a gene such that, no matter what, they will turn out to be gay. Another, larger group, say 10%, might have a gene such that they could be gay, depending on their environment. For such people, their sexual orientation would be determined by a single gene allele. Same-sex attraction in such people will be caused by a combination of several genes, and the interaction of this combination with a particular environment. Finally, there will be some people who lack either gene but who just decide to be gay. (Thanks to Udo Schuklenk for help on this point.)

22. Barbara Kantrowitz, "Gay Families Come Out," *Newsweek*, 4 November 1997, 53.

23. For a description of this testimony, see the Internet site gaytoday@badpuppy .com and the story, "Clone Rights Advocates Scold Scientists."

Arguments against Human Asexual Reproduction

And the Lord God caused a deep sleep to fall upon Adam, and he slept; and he took one of his ribs, and closed up the flesh instead thereof;

And [from] the rib, which the Lord God had taken from the man, made he a woman, and brought her unto the man.

And Adam said, This is now bone of my bones, and flesh of my flesh.

Genesis, 2: 21–23

· · · · ·

In this chapter, I examine arguments claiming that human asexual reproduction is intrinsically immoral, as well as arguments that claim that human NST is not directly immoral, but that its indirect effects warrant its prohibition. I shall argue that neither kind of argument justifies a reproductive ban of this kind, especially if asexual reproduction is limited to cases where it will likely produce benefits to the child and to its parents.

Against the Will of God

The most general argument against creating an adult human being by NST is that it is against the will of God. Some people believe that humans have no right to change the way that humans are created because sexual reproduction was ordained by God and that to try to change this way is sinful.

Upon hearing the news of Dolly, Duke University divinity professor Stanley Hauerwas said that those who wanted to clone Dolly "are going to try to sell it with wonderful benefits" to medical and animal industries. He condemned the procedure because he thought it was "a kind of drive behind this for us to be our own creators," raising the images of Drs. Frankenstein and human hubris. A scant ten days after Dolly's announcement, the Christian Life Commission of the Southern Baptist Convention predictably called for a federal law against human cloning, as well as an international law to the same effect.[1] The Vatican in April of 1997 asserted: "the right to be born in a human way." It urged all nations to ban human cloning and, in manner typical of Vatican ethics, urged them to make no exceptions.[2] An enlightened Shi'ite Muslim jurist named Sheikh Fadlallah suggested that originating a child by cloning should be punished by death or, at the very least, amputation.[3]

The NBAC *Report* quoted Paul Ramsey's dictum that, "Religious people have never denied, indeed they affirm, that God means to kill us all in the end, and in the end he is going to succeed."[4] As said, testifying before NBAC, Lutheran theologian Gilbert Meilaender emphasized the importance of "the creation story in the first chapter of Genesis [that] depicts the creation of humankind as male and female, sexually differentiated and enjoined by God's grace to sustain human life through procreation."[5] Other theologians testifying emphasized the Scriptural basis of: God's creation of humans, warnings not to play God, dangers of the quest for knowledge, the need for responsible dominion over nature, and the ultimate folly of human destiny.[6]

Verses 26–27 of the first chapter of Genesis tell us, "And God said, Let us make man in our image, after our likeness . . . So God created man in his own image, in the image of God created he them; male and female created he them." Two chapters later, we get another famous, relevant story:

6 And when the woman saw that the tree was good for food, and that it was pleasant to the eyes, and a tree to be desired to make one wise, she took of the fruit thereof, and did eat, and gave also unto her husband with her; and he did eat.

7 And the eyes of them both were opened, and they knew that they were naked; and they sewed fig leaves together, and made themselves aprons

This story tells an oft-told tale in the Old Testament that once-upon-a-time there was a time of moral perfection, but because of some bad act of humans, things changed and disaster followed. Genesis describes Eden as a pure and simple land that ultimately gets so corrupted that God brings the great flood, allowing only Noah and his family to escape. Often the agent of corruption is the human quest for knowledge, represented by the Tower of Babel, which also signifies human hubris before God.

These stories are our most ancient tales and, as such, resonate deeply within us. They convey the message that a small bit of new knowledge can be treacherous for humanity. Humanity is forbidden from knowing certain things, and the consequence of knowing these things is that a black cloak of doom will cover our world.

Despite these deeply-ingrained feelings, it is a grave mistake to let such ancient stories dictate modern public policy. First, and most obviously, they are just stories. Many of the events described in Genesis, especially the story of the Garden of Eden, are not actual events but symbolic fables, repeated in many Near Eastern cultures. For example, pre-Hebraic civilizations such as Babylonia also had the story of a great flood.

Second, and of equal importance, the Old Testament is not full of moral wisdom. Although Meilaender's entire testimony derives from these Scriptural stories, like all such references to Scripture, he only goes there to justify what he already believes. If a neutral person merely read the Old Testament and followed what it said, he would find that: Abraham and King David make women captured in war into slaves in their harems; Exodus 21:2 condones human slavery; Jephthah's daughter is required as a human sacrifice; sex with slave-maids is permitted to create male heirs when Sarah's and Rachel's husbands are barren; and Exodus 23:27 permits killing women and children of enemies in war.

Genesis 3:16 also says famously, "Unto the woman he said, I will greatly multiply thy sorrow and thy conception; in sorrow thou shalt bring forth children; and thy desire shall be to thy husband, and he shall rule over thee." Obviously, this event could not have really happened and must just be a story, because otherwise we are forced to see God as utterly unjust. Not only is Eve punished for her own act, so are billions of women afterwards in human history who must experience painful childbirth. Although Scripture condones these injustices, Meilaender undoubtedly does not think them right, and that is because he has some non-Scriptural standard of right on which to base his judgment. It is also worth noting that rabbis such as Moses Tendler, who also testified before NBAC and whose religion has been studying Genesis longer than Christianity, do not think that the creation story of Genesis implies an absolute ban on human cloning.

Consider also that the story of the Tree of Knowledge is about moral knowledge: "then your eyes shall be opened, and ye shall be as gods, knowing good and evil" (3:5). What should we infer from this story. That humans should know nothing about morality? Perhaps humans made no moral judgments in the Garden of Eden, but how is that story relevant to a world where humans must constantly do so?

To accept Old Testament morality in general is to accept a fatalistic worldview. That is understandable because the Old Testament was edited to its final form over two thousand years ago when being fatalistic made sense. The Old Testament did not anticipate genetic therapy, artificial skin, and organ transplants.

In more modern theology, it takes a mighty lot of intermediate steps of interpretation and reasoning to get from (1) A rational God exists and cares about us, to (2) It is immoral to originate a human by NST. Such steps would have to counter the premises that, (3) Because a rational God exists and cares about us, He allows us to make new discoveries in medicine and science, and that, (4) Because God is rational and cares about us, he directs us to create humans in ways that are rational and that express caring about human beings. Substantial argument would be needed to prove that originating a child by NST by parents with good motives was against (4). I do not think that can be done.

After all the dust settles about originating babies by NST, it may turn out that opposition to it stems *only* from religious people or from those whose thinking is colored by traditional religious beliefs. If so, isn't there a problem with recognizing such views in American law? The U.S. Constitution forbids the establishment of laws and policies that are solely motivated by religious beliefs. Should studies in non-human mammalian embryogenesis indicate that NST would be safe in humans, and should well-motivated parents be identified who had good reasons to try it (being at risk for having babies with genetic disease), would not further opposition be solely based on religious grounds? And as such, would it not violate our Constitution to have it ground a federal ban that restricts others' reproductive freedom?

Fear of the New and Different

In his testimony before the Senate, George Annas opposed NST in humans because "it would radically alter the very definition of a human being by producing the world's first human with a single genetic parent."[7] Of course, the whole meaning of "parent" these days is ambiguous. Many people understood a similar point when surrogate "mothers" became possible: there was the egg donor (aka the biological mother), the gestational mother, and the mother who raised the child (for some, merely the social mother; for others, the real mother).

Originating a child by NST now gives males a chance to participate in this ambiguity because a male can be a "parent" as a sole genetic source, as a donor of sperm, or as a person who raises the child. Women can now be "mothers" in four senses.

All this means that Annas is wrong. A child created by NST would not have "a single genetic parent" but four "genetic" parents plus a gestational mother and possibly a social mother. Regardless of what one thinks of having so many parents, the problem is certainly not having only one parent. (Lori Andrews, a law professor in Chicago, also thinks that the law might initially consider the source of the genes of the baby as the legal parents, especially if DNA testing is used.[8])

Annas' worry about crossing natural boundaries would be rational if, and only if, some new kind of genetic material were involved (which it is not) or if genetic material from another species was involved. But the engines of making a human are the same old human genes that always have made humans. Biologically, the parents of a child originated from Greg Pence would not be Greg and Pat Pence (my wife and I) but Gilbert and Louise Pence (Greg's parents).

Nevertheless, behind Annas' objection lies a more primitive, pervasive fear, that goes like this: Human heterosexual reproduction is a distinct kind of reproduction, like a species, and asexual reproduction is of another kind, such that the move from one to the other has unknown dangers, is going against nature, and exhibits human hubris.

The worry expressed here is actually a cluster of related worries that all revolve around fear of a change about something new. The worry can be put in many different forms:

1. NST is too radical and too different to try.
2. Things are changing too fast in medicine, and NST is another example. Better to keep to the tried-and-true ways.
3. The "Jurassic Park" objection—If we try to mess with changing something so primordial as human sexual reproduction, something will go wrong and "bite" us.
4. NST is replication, manufacture, artificial, and as such, reduces human specialness.
5. It is hubris to tempt fate by mucking around with human nature.

I try to answer such worries in the following sub-sections.

Change Itself Is Neither Good Nor Bad In Itself
Every big change in medicine at first worries people, as people worried when physicians first transplanted hearts or when human gene therapy was first tried.[9] "Biological innovations are often regarded as perversions when they first come along," according to California Institute of Technology's Daniel Kevles, a famous historian of eugenics. "Then they become widely practiced."[10] Medical sociologist James Hughes said, "I'm uncomfortable

with sanctimonious and ill-thought-out presidential decrees where the President has determined that we have crossed a boundary beyond which humanity dare not go. I disagree that there is such a boundary. Humanity should be in control of its own destiny."[11]

There is nothing about change itself that is bad. For example, a drug may soon be available that increases memory in humans by 10%. While it would certainly cross a natural boundary to allow adults to use it, it would not be wrong to do so just because humans have always been limited in this way. Putting a pig liver into a human also crosses a natural boundary, but if my sister is dying of liver disease and it will give her another decade of life, I want her to have it.

In general, the reaction to the cloning of Dolly revealed that many people are fatalistic, even though they might not use this word to describe their view. Thousands of letters and comments in the media expressed the general view that new developments in reproductive medicine "have gone too far." This widely-held view is part of a more general, fatalistic view: Man does not exist to control Nature; man exists within Nature. New techniques of assisted reproduction create a false, wrong sense that human life is something within our control.

Rabbi Marc Gellman of Temple Beth Torah in Melville, New York, aptly summed up this view:

There is a wisdom in common people, in ordinary people who have been unfairly demeaned by people who view their lack of knowledge about haploid and diploid and totipotent embryos as somehow a disqualification to have moral sensibilities . . . There is a strong and real understanding that we are not our own creators. And this [cloning] technology undermines that fundamental belief in the most powerful and disturbing way possible.[12]

According to this view, we do not own ourselves. We cannot create, reproduce, and kill ourselves as we wish. Theologian Paul Ramsey championed this view in the 1970s.[13] He argued that human life was a gift from God and that one does not throw back such a gift in the face of the Giver by taking one's own life.

A frequent corollary to the fatalist viewpoint is that human nature is not to be trusted with any new knowledge. Any attempt to change our natures will produce dark consequences for humanity, often by humanity itself. For Ramsey and countless others, any attempt to master human nature means some men mastering other men, which really means the enslavement of some men by others.

Fatalism has many variations. One extreme is the position that we should never change anything. An even more extreme position might be that of Luddites and nature romanticists, who would return us to a simpler, purer time (but of course that would involve changing the present, which true fatalists oppose). The more common fatalistic view is that everything is changing too fast, especially sex roles and reproduction.

Father Richard McCormick said that the origination of a person by NST seems to be closer to designing people "to our specifications," an enterprise that he likened to trying to be like God.[14] Such origination, he said, would tempt people to create persons with certain kinds of physical and intellectual traits, elevating mere aspects of being human above "the beautiful whole that is the human person."

Christian apologist C. S. Lewis expressed an essentially fatalistic view when he asked whether we would be able to get back to the unmanufactured if we decided that we didn't like man-made man.[15] For Lewis, we shouldn't manipulate man and try to improve on nature.

The voluntarist opposes the fatalist. For him, we have the wisdom to use new knowledge to help ourselves. Knowledge in itself is neither good nor evil. Knowledge is knowledge. It is human choice that is good or bad, moral or immoral.

So humanity must choose to go forward. Not to choose is a choice, and to choose not to go forward is to choose to stagnate. Many times in the past, societies have turned backwards and away from change. As a result, they died out or became but shells of their former, vigorous selves.

For this author, the point to emphasize here is that every improvement in medicine is anti-fatalistic. Almost everything in medicine—from rehabilitative medicine to neonatal intensive care units to preventive geriatric medicine—opposes surrender to the primitive and allegedly ordained. Only hospice care at the very end of life is really fatalistic.

One of the pioneers of modern bioethics, and a champion of voluntarism, was the late Joseph Fletcher, a frequent opponent of Paul Ramsey, Leon Kass, and other intellectuals aligned with fatalism. "Old Joe" foresaw many of the problems we face today and foresaw the backwards-looking answers that many of our public moral voices would give:

> The real choice is between accidental or random reproduction and rationally willed or chosen reproduction. . . . Laboratory reproduction is radically human compared to conception by ordinary heterosexual intercourse. It is willed, chosen, purposed and controlled, and surely those are among the traits that distinguish Homo sapiens from others in the animal genus, from the primates on down.[16]

The fatalist does not think that life is, or should be, within human control. Physician-assisted suicide, abortion, contraception, and assisted reproduction for fatalists are all unnatural, and since "unnatural" for them equals "unethical," these practices should be banned by ethics, public policy, and law.

Yet fatalists always equivocate about nature. They romanticize the past, "the way things were," seeing it through the rose-tinged glasses of selective attention. They are like those who believe in reincarnation and believe themselves to have been an Egyptian prince, never the common dirty Egyptian slave with bad bowels and no teeth.

Nature is not always great. As Tennyson said, real nature is "red in tooth and claw." Defenders of the social status quo exaggerate the extent to which the natural is desirable. Actually, the status quo which they defend is not natural at all, but merely the man-made form of the artificial society that they understand, i.e., a social artifact they want frozen as if it were human nature itself.

Another thing to be said for voluntarism is that it properly identifies the sources of what is good in human civilization. In contrast to pedigreed dogs, cats, horses, and cattle, very little has been contributed to the last five thousand years of human civilization by the transfer of genetic information. Instead, it has been the transfer of non-genetic information that is the essence of civilization: the transfer of the ability to read and do math through basic education, the transfer of the research of previous generations as knowledge contained in books and computers, and the transfer of ways of growing food, ordering human societies, making peace, and so on that constitute who we are. None of these things are really threatened by human asexual reproduction.

As English philosopher Jonathan Glover notes in *What Sort of People Should There Be?*, it is not as if by banning human asexual reproduction we would not be allowing genetic change.[17] Clean water, better nutrition, safe housing, and medical care allow people to pass on their genes who previously might have died early. Jets allow rapid transportation around the globe and new kinds of marriages. The global economy moves people in and out of countries. Immigration and emigration constantly occur between countries. Radiation, drugs, and human exposure to chemicals change the genes that are in us, including our germ cells. We are slowly moving to a future where everyone has some shade of brown skin.

Ultimately, I think, most worries about change from NST stem from some conservative, religious viewpoint, even though the language may shift from religious terms to more neutral ones about Nature, human nature, and society (or even, rates of change, community and stability). The basic intuition is that certain things are, and should be, fixed about people, and medicine shouldn't change those things. As I have argued, when it comes to eradicating disease and improving performance, medicine always rejects this view.

Not So New and Different

A different line of reasoning is to say that NST isn't really so new or different. Consider some cases on a continuum. In the first, the human embryo (call the first existing female embryo, Rebecca) naturally splits in the process of twinning and produces two genetically-identical twins. Mothers have been conceiving and gestating human twins for all of human history. Call the children who result from this process Rebecca and Susan.

In the second case a technique is used that could be used in infertility clinics where a human embryo is deliberately twinned in order to create more embryos for implantation in a woman who has been infertile with her mate. (This is what infertility researcher Jerry Hall did in 1993.) Instead of a random quirk in the uterus, now a physician and an infertile couple have chosen to introduce a tiny electric current to split the embryo. Two identical embryos are created. All embryos are implanted and, as sometimes happens, rather than no embryo implanting successfully or only one, both embryos implant. Again, Rebecca and Susan are born.

In the third case, one of the twinned embryos is frozen (Susan) along with other embryos from the couple and the other embryo is implanted. In this case, although several embryos were implanted, only the one destined to be Rebecca is successful. Again, Rebecca is born.

Two years pass, and the couple desire another child. Some of their frozen embryos are thawed and implanted in the mother. The couple knows that one of the implanted embryos is the twin of Rebecca. In this second round of reproductive assistance, the embryo destined to be Susan successfully implants and a twin is born. Now Susan and Rebecca exist as twins, but born two years apart. Susan is the delayed twin of Rebecca.

(Rumors abound among physicians that such births have already occurred in American infertility clinics. The author himself knows of one such couple where a delayed twin birth probably has occurred in a couple that used several cycles of IVF to produce two children; this couple has chosen so far not to prove whether the two siblings are identical and to keep the news of their children private.)

Suppose now that the "embryo that could become Susan" was twinned, and the "non-Susan" embryo is frozen. The rest of the details are then the same as the last scenario, but now two more years pass and the previously-frozen embryo is now implanted, gestated, and born. Susan and Rebecca now have another identical sister, Samantha. They would be identical triplets, born two and four years apart. In contrast to NST, where the mother's contribution of mitochondrial genes introduces small variations in nearly-identical genotypes, these embryos would have identical genomes.

All these events occur because people deliberately exercise choice and control over a natural process that could leave them otherwise barren and frustrated. These artificial processes result in children who are very much desired and wanted, not an inconsiderable asset with which to be born in these days of indifferent parenting.

Next, suppose that the embryo that could have been Rebecca miscarried and never became a child. The twinned embryo that could become Susan still exists. So the parents implant this embryo and Susan is born. Query to National Bioethics Advisory Commission: have the parents done something illegal? A child has been born who was originated by reproducing an embryo with a unique genotype. Remember, the embryo-that-could-become Rebecca existed first. So Susan only exists as a "clone" of the non-existent Rebecca.

Now, as bioethicist Leroy Walters emphasizes, let us consider an even thornier but more probable scenario.[18] Suppose we took the embryo-that-could-become Susan and transferred its nucleus to an enucleated egg of Susan's mother. Call her "Suzette", because she is like Susan but different, with new mitochondrial DNA. Now the "Susan" embryo was created sexually but Suzette's origins are through nuclear transfer. It is not clear that this process is illegal. The NBAC *Report* avoids taking a stand on this kind of case.

Finally, we could compare all the above cases to originating Susan by NST from the genotype of the adult Rebecca. Susan would again have a nearly-identical genome with Rebecca. Here we also have nearly identical female genotypes, separated in time, created by deliberate choice with the aid of physicians specializing in infertility assistance. Originating a child by NST can be seen as not a breakthrough in kind but as a matter of degree.

Consider human assisted reproduction. A woman is given drugs to stimulate superovulation so that surgeons can remove as many eggs as possible. At each cycle of attempted in vitro fertilization, 3 or 4 embryos are implanted. Most couples make several attempts, so as many as 9 to 12 embryos are involved.

Unfortunately, most couples do not take home a baby (less than one in five). This means that for every baby born by in vitro fertilization, as many as a dozen embryos were implanted and failed to gestate. Seen in this light, what Wilmut did, starting out with a large number of embryos to get one successful lamb at birth, is not so novel or different than what now occurs in human assisted reproduction. And remember 40% of embryos fail to implant.[19] Human attempts at NST should certainly not be held to a higher standard than normal, assisted, human sexual reproduction.

We Are Already Doing More Radical Things than NST

The above line of reasoning argued that NST would not be really so new. A related line of argument is to note that we are already doing things far more radical than human NST.

For example, putting human genes in pigs to create possible organ transplants from such altered pigs is far more radical than human NST. Transplants from such pig organs open up the possibility of a two-way travel of porcine viruses to humans and vice-versa (of concern to those who think that AIDS came from simian-human contact).

Speaking of animals, we crossed a natural barrier a decade ago when we allowed Harvard University in 1987 to patent its oncomouse (aka the "Harvard oncomouse"). Since then over a thousand applications for such patents have been filed and over 50 patents on genetically-altered or genetically-created animals have been issued by the U.S. Patent Office.[20] (Remember, Dolly was born in July of 1996; Wilmut and PPL Therapeutics waited until February 24, 1997 to make their announcement because only at that time did they have a protected patent.) Overall, from genetically-altered tomatoes to pig-grown livers for transplanting into humans, we are doing radically new things to save human lives, crossing "natural" barriers all the time, and hardly blinking an eye about it.

Genetic Diversity and Evolution

A different objection to NST is that it will decrease the diversity of the gene pool. Diversity is a good thing in any population. If a population becomes too in-bred, it loses the ability to respond to a new threat, such as a lethal virus. This would be a problem for the animal industry if one kind of lamb or cow was mass-cloned for products of use to humans. In humans, a scarcity of food or of fertile mates, might call for new skills only available in a population with a very diverse gene pool. The basic objection is that you never know what genes you're going to need, so we shouldn't mess around creating and indirectly eliminating some genes.

The great problem with this argument is that it commits the all-or-nothing fallacy: either all human reproduction is going to be sexual reproduction or none is. Appeals to the livestock and plant industry backfire here. The most common practice in the animal industry is to mix sexual and asexual reproduction to guarantee genetic diversity. Sexual reproduction is used experimentally to try to discover better kinds of animals and plants, but in the meantime, superior, healthy animals and plants are reproduced asexually.[21]

Moreover, and as NIH Director Harold Varmus said in testifying before Congress, originating children by NST is never going to be a widespread option.[22] The costs and rates of failure that typify in vitro fertilization—the only way to do NST now—are not something that most couples will prefer, even if they have concerns about familial genetic disease.

What this means, practically, is that were a few hundred thousand couples in advanced countries to gestate children by NST, it would have no impact on the planetary human genome. Indeed, by the time such couples can so gestate, the planet will likely have not six billion but eight billion humans, and so the force of the "diversity objection" reduces to a vanishing point.

But let's give the argument a little rope and see if it hangs itself. Suppose that millions of humans one day are created by NST. Is that going to eliminate any potentially useful genes from the human gene pool?

When faced with a similar argument, early twentieth century geneticist Herman Muller once retorted that, "An ounce of algebra" was worth "a ton" of ideology. A primary law of population genetics is regression to the mean. If a society deliberately tried to decrease diversity in the human population, it would be a hard task. Most of the other six billion humans will still be at work passing on their genes, and that means that powerful forces will preserve the human genome. Even the most vast attempts to improve the human genome will not get around the inherent tendency of the human population to revert to the normal range.

Take it one step further, to its most ridiculous assumption. Suppose that all the governments in the world tried assiduously to improve the human genome. Even so, it is doubtful that they could. Human reproduction is a notoriously difficult thing to control. Boy meets girl once in the night and a child is conceived. Despite what governments tried to do, the reality is that the billions of existing people would keep on creating more people in the old way, mixing their genes with people originated by NST, with everyone ratcheting down to the mean.

More practically, for every high-minded couple who produced a superior child by NST there would be a Brazilian couple who produced nine children by normal sex. As people find it easier and easier to migrate and travel around the world, genes will be cross-fertilized, making regression to the mean even faster. In short, the genetic diversity of the genome is not threatened by human NST.

Behind the argument for genetic diversity is sometimes the idea that humans should not stop evolution and that human NST would do so. One problem here is that it is a very big leap in ethics to say that human evolution is good and that therefore, whatever furthers human evolution is also good.

English philosopher Mary Midgley observes that some people trade worship of a deity for worship of evolution.[23] Perhaps her most devastating criticism of such people's beliefs is that evolution should never be seen as a pyramid with humans at the apex. It is more accurately seen as a bush with humans at the top of one branch somewhere in the upper half, but also with many other branches in that upper half. What species emerges as the next tallest branch will depend on many factors, including whether humans destroy themselves.

Human NST will not end evolution. It does not substitute man-made genes for those inherited from our ancestors. Human asexual reproduction is not a threat to the ends of evolution.

Risk of Harm to the Child

The chief practical and moral objection to human asexual reproduction is that a child might be harmed by NST. This is such an important objection that I discuss it in two different forms: (1) risk of harm from mistakes in genetic transfer or fetal development, and (2) risk of harm from unrealistic expectations. The first objection focuses on risks of the baby being a monster or having a gross physical defect; the second focuses on psychological harm from social expectations.

(I) Mistakes in Genetic Transfer or Fetal Development
Will a child originated by asexual reproduction be at more risk of being abnormal at birth than a child with two genetic sources? There are three known possible things that could go wrong in nuclear somatic transfer to produce a human child: (1) the parental imprinting of DNA could go wrong as the DNA gets modified by methylation; (2) the chromosomes could get re-modeled (this could explain Wilmut's low percentage of live births from the 277 sheep embryos); (3) the telemeres (the ends of chromosomes) shorten as mammals age; if this is passed on by nuclear transfer, it could affect the genome of the child and shorten the expected life of the child.[24] It is also theoretically possible, although unlikely, that something totally unknown could go wrong.

Many critics incorrectly assume that a child originated by NST would be in danger of greater genetic or physical injury than the average human child. There is one piece of evidence against this view: we already know the NST child's genetic makeup: we know its adult nature and that it made it through gestation. On the other hand, if we started from scratch with sexual reproduction, we know that of the 40% of human embryos that won't implant, half don't because of a genetic defect.[25] With an NST child, we know that its embryo has already been successful in implanting, and hence, that it has a higher-than-normal chance of achieving fetushood. Similarly, we know that it did not miscarry because of some genetic flaw. We know that as a child it did not develop a hitherto-unsuspected genetic disease. Hence, we may not be subjecting the cloned embryo to greater, but to lesser, risk of genetic injury than normal gestation.

Balanced against the risks of asexual human reproduction are the known risks of human sexual reproduction, which include: (1) being born with too many or too few chromosomes, (2) having both parents be carriers for a genetic disease and giving such a disease to a newborn, (3) having a gene mutation occur that causes a deleterious defect. Human asexual reproduction would run almost none of these risks.

At this moment in time, animal tests have not shown that NST is safe enough to try in humans, and extensive animal testing should be done over the next few years. That means that, before we attempt NST in humans, we will need to be able to routinely produce healthy offspring by NST in lambs, cattle, and especially, non-human primates.

After this testing is done, the time will come when a crucial question must be answered: how safe must human NST be before it is allowed? This is probably the most important, practical question of this book.

Should we have a very high standard, such that we take virtually no risk with a NST child? As explained in Chapter 3, critics in the past (such as Daniel Callahan and Paul Ramsey) implied that unless a healthy baby can be guaranteed the first time, it is unethical to try to produce babies in a new way. At the other extreme, a low standard would allow great risks.

What is the appropriate standard? How high should be the bar over which scientists must be made to jump before they are allowed to try to originate a NST child?

In my opinion, the standard of Callahan and Ramsey is too high. In reality, only God can meet that Olympian standard. It is also too high for those physicians trying to help infertile couples. If this high standard had been imposed on these people in the past, no new form of assisted reproduction would ever have been tried.

On the other end of the scale, one could look at the very worst conditions for human gestation, where mothers are drug-dependent during pregnancy or exposed to dangerous chemicals. These conditions should also include parents with a 50% chance of passing on a lethal genetic disease to their offspring. The lowest standard of harm allows human reproduction even if there is a 50% chance of harm to the child ("harm" in the sense that the child would likely have a sub-normal future.) One could argue that since we "allow" such mothers and couples to reproduce sexually, we could do no worse when originating by NST.

I believe that the low standard is inappropriate to use with human NST. There is no reason to "justify down" to the very worst conditions under which society now tolerates humans being born. If the best we can do by NST is to produce children "as good as" those born with fetal-maternal alcohol syndrome, we certainly shouldn't originate by NST.

In between these standards, there is the normal range of risk that is accepted by ordinary people in sexual reproduction. I believe that human NST should be attempted when the predicted risk from animal studies is no greater than such a normal range. By "ordinary people" I mean those for whom the mother is neither an alcoholic nor dependent on an illegal drug and where neither member of the couple knowingly passes on a high risk for a serious genetic disease.

This standard seems reasonable to me. It does not require a guarantee of a perfect baby, but it also allows not just anything to be tried. For example, if the rate of serious deformities in normal human reproduction is 1–2%, and if the rate of chimpanzee NST reproduction was brought down to this rate, and if there was no reason to think that NST in human primates would be any higher, then I think it would be permissible to attempt human NST.

There is one line of reasoning that defends the low standard that I wish to mention and then dismiss. It says a child cannot be harmed by being brought into the world by a method where, without it, the child would not exist. According to this view, any existence is better than non-existence.[26] Even with paralysis, who can complain if the alternative is not to exist at all? Indeed, because the alternative is non-existence, which itself does not exist, there cannot even be a plausible comparison here of the paralyzed child's life, for there is nothing to compare the impaired child's life with.

The problem with this line of reasoning is that it justifies doing almost anything to the fetus. Short of a wrongful life of total pain, any existence can be argued to be better than non-existence. True, there is no alternative child waiting to be born in the wings of the theater of existence, who might be created if this child were not, because in the scenarios we are discussing, the only way for this child to be born is through the technique that might harm him. Nevertheless, I don't think this justifies using the low standard.

One great confusion here stems from not distinguishing among different kinds of harm. When we jump from considering this particular child to a whole class of children who might be born, then we see that some children can be plausibly said to be harmed if they are less than normal. If someone tried out risky new drugs on fetuses, but on fetuses which otherwise would have been aborted, and with the result that some were born as babies missing arms and legs, it would not be correct to say that nothing wrong was done because these particular babies could not have been harmed because otherwise they would not have existed. Instead, we would judge that harm occurred because this class of babies did not need to exist and was caused to exist at an inferior quality of life by the researchers allowing harm to occur which need not have occurred. So I do not think this is a good defense of the low standard.

I want to make one more point about just how high the average standard must be. One factor increasing fear of NST is its very novelty. However, in more familiar contexts and with more familiar motives, we allow much greater risks to intended children. Old common dangers, such as women who have children as teenagers, women who drink and smoke during pregnancy, or women whose mates work with dangerous chemicals, are seen as less threatening than brand-new dangers from asexual reproduction

In the year or two before Dolly's birth, IVF clinics had a rapid and uncritical acceptance of intracytoplasmic sperm injection (ICSI), the practice of using one sperm to impregnate an egg for men with very low sperm counts. Such a practice, one defender of NST noted, "was widely made available at a time when experimental evidence as to its [ICSI's] safety was still flimsy."[27] Because we accept as normal the desire of heterosexual couples to have children with their genes, no outcry was raised over the introduction of this technique. Indeed, in the "Science Times" section of the *New York Times*, ICSI was discussed in the tone of it's-another-miracle-in-medicine.

The point is not that two wrongs make a right, but rather, that we tolerate a good deal of risk to the unborn in situations that are familiar. It may be mainly the novelty of NST that makes us fear its unknown risks. Once the risks of NST become more familiar and predictable, we may be able to think about it more rationally.

Up until now, many people seem to have exaggerated the possible risks of NST. As my colleague G. Lynn Stephens remarked about the things people were saying about human cloning after Dolly's birth, it is as if people had said upon seeing the first car, "Gee, those things ought to be banned because people might use them to run over each other."

(II) Risks of Harm from Unrealistic Parental Expectations

There is a widespread belief that any child originated by NST would be harmed by unrealistic expectations created by her parents in comparing her future to the life of her genetic ancestor. A lot of weight may rest on the strength of this objection.

In giving an overview of NBAC's reasons for rejecting human cloning, law professor Alex Capron (and NBAC member) said that NBAC's conclusions to ban human NST really rested on "two pillars": (1) the safety issue (physical harm to the child) and (2) the expectations issue (whether unrealistic parental expectations might harm the child).[28] (Other worries, such as harm to the family or to society, he said, were seen as nebulous and speculative.)

Capron's overview becomes more important when linked to comments by many scientists and bioethicists at the same meeting that the safety issue would vanish in a few years as results accumulated from studies of non-human, mammalian NST. This means that within a few years, all the moral weight to human NST could rest on the argument about expectations. (This also means that the Commission proposed laws to ban human NST, even though it thought such a procedure might one day soon be made safe for children.)

According to the objection about expectations, choosing to have a child is not like choosing a car or house. It is a moral decision because another being is affected. Too many immature people have children—the phenomenon of "children having children"—and that is not good for the babies. Having a child should be a careful, responsible choice and focused on what's best for the child. Therefore, having a child originated by NST is not morally permissible because it is not best for the child.

The problem with this argument is the last six words of the last sentence, which assumes bad motives on the part of parents. Unfortunately, NST ("cloning") is associated with bad motives in our culture, but until we have evidence that it will be used this way, why should we assume the worst about people?

Certainly, if someone deliberately brought a child into the world with the intention of causing it harm, that would be immoral. Unfortunately, the concept of harm is a continuum and some people have very high standards, such that not providing a child a stay-at-home parent constitutes "harming" the child. But there is nothing about NST per se that is necessarily linked to bad motives. True, people would have certain expectations of a child created by NST, but parents-to-be already have certain expectations about children.

I advise very high-achieving students in a college program where they are accepted (so long as they complete a standard premedical curriculum) directly out of high school into medical school. These students are very bright, talented, and personable. In almost every case, they have parents who have very high expectations of them and who are very involved in their daily lives. The expectations of these parents have produced some of the best students in our university.

Another version of this argument is that it is arrogant for a parent to create a specific type of human life. Just as it is hubris for a parent to take her two-year old girl to four hours of dance a day to try to force her to be a great dancer, so the parent who originates the genes of a great ballerina misses the central point about being a parent. The child should be valued for who she is, not what she could become. It is wrong to see a child as a lump of clay to be molded by a parent's expectations.

In reply it can be argued that too many parents are fatalistic and just accept whatever life throws at them. The very fact of being a parent for many people is something they must accept (because abortion was not a real option). Part of this acceptance is to just accept whatever genetic combination comes at birth from the random assortment of genes.

But I fail to see why such acceptance is a good thing. It is a defeatist attitude in medicine against disease; it is a defeatist attitude toward survival when one's culture or country is under attack; and it is a defeatist attitude toward life in general.

Daring to expect greatness of one's children seems like a much better attitude. True, parents with high expectations will be very disappointed with normal achievement, but such parents often take extraordinary precautions to maximize the potential of their kids, e.g., by sacrificing to get them the very best education. Originating a child by NST gives parents the chance to carry on the genes of people they admire.

Indeed, most of us would be happy if our parents had created us by attempting to give us a genetic inheritance free of cancer, heart disease, and progressive neurological disease by originating us from some person known to have lived into her nineties with an exuberance for life. If we applied the Golden Rule, it is reasonable to assume that—far from being depressed at such origination—most people would be glad to have been originated from the genotype of someone we now admire.

Peter Singer, the Australian bioethicist who started the animal liberation movement two decades ago and who has written widely about issues of life and death, considered (with Dean Wells) the possibility in 1984 that this might put too much psychological pressure on a child:

> Parents who had gone to considerable trouble to obtain a clone of a
> brilliant scientist or leading statesman would not find it easy to hide

their disappointment in a child who lacked the qualities they thought to guarantee. Would they be able to love their children with the uncritical love of parents who accept their children for what they are—a love which does not depend on the extent to which the children measure up to some preconceived standard of excellence? And will the children not suffer from the knowledge that they are not living up to their parents' expectations? Of course, many children know they are not living up to their parents' expectations; but the impact would be immeasurably greater if the child knew that his parents had gone to the trouble of ensuring that it had all the genetic characteristics necessary for the expected, but unattained, level of achievement.[29]

But is there really a reason to expect that children will be harmed by parents' expectations? Why not wait and see whether parents adjust when their expectations are disappointed? As Singer and Wells observed, parents who expect a boy quickly adjust to having a girl and may come to dote on the girl. Moreover, is this not the old fallacy of assuming bad motives in parents? Why assume that parents who originate a child by NST will be less able to adjust their expectations? In the back of our minds, are we thinking of pushy tennis mothers? But such mothers are only the tiniest of fractions of ordinary mothers.

One reason why some parents are disappointed stems from the random pattern of gene assortment. It is true that two parent-musicians would certainly be disappointed if their child was tone-deaf, but how likely is that if the ancestor has perfect pitch?

Leon Kass objects that defenders of human cloning speak out of both sides of their mouth. On the one hand, they say that a child originated by NST from Michael Jordan wouldn't be exactly like Michael Jordan, would have the free will to choose not to play basketball, and have a different upbringing, such that he might choose to become an evangelical minister (and hence, expectations about children originated this way might be simplistic). On the other hand (and here is the only chuckle in his essay):

But one is shortchanging the truth by emphasizing the additional importance of the intrauterine environment, rearing and social setting: genotype obviously matters plenty. That, after all, is the only reason to clone, whether human beings or sheep. The odds that clones of Wilt Chamberlain will play in the NBA are, I submit, infinitely greater than they are for clones of Robert Reich.[30]

Well, yes, of course. As previously argued, there will always be some reason for choosing to originate a child from this genotype rather than that one.

But it is a big jump, and not a justified one, to claim that because a parent tries to choose a better genetic future for a child than what might be obtained from random genetic combinations, that the parent will only be satisfied with a perfect child.

We also have the old fear of bad motives lurking here. "The expectations of parents will be too high!" it is asserted. "Better to leave parents in ignorance and their children as randomness decrees." The silliness of that view is apparent as soon as it is made explicit.

If we are thinking about harm to the child, an objection that comes up repeatedly might be called the argument for an open future. "In the case of cloning," it is objected, "the expectations are very specifically tied to the life of another person. So in a sense, the child's future is denied to him because he will be expected to be like his ancestor. But part of the wonder of having children is surprise at how they turn out. As such, some indeterminacy should remain a part of childhood. Human NST deprives a person of an open future because when we know how his previous twin lived, we will know how the new child will live."

It is true that the adults choosing this genotype rather than that one must have some expectations. As said, there has to be some reason for choosing one genotype over another. But these expectations are only half based in fact. As I argued extensively in the first chapters, no person originated by NST will be identical to his ancestor because of mitochondrial DNA, because of his different gestation, because of his different parents, because of his different time in history, and perhaps, because of his different country and culture. To assume that such a child's future is not open is incorrectly to assume naive genetic reductionism.

Moreover, notice that almost all the expected harm to the child stems from the predicted, prejudicial attitudes of other people to the NST child. ("Would you want to be a cloned child? Can you imagine being called a freak and having only one genetic parent?")

As such, it is important to remember that social expectations are merely social expectations. Social expectations are malleable and can quickly change. True, parents might initially have expectations that are too high and other people might regard such children with prejudice. But just as such inappropriate attitudes faded after the first cases of in vitro fertilization, so they would fade here too.

Ron James, the Scottish millionaire who funded much of Ian Wilmut's research, points out that before the announcement of Dolly, polls showed that people thought the following to be morally problematic: cloning animals, gene transfer to animals; and germ-line gene therapy was thought to be just wrong. In only two months after the announcement of Dolly, and after much discussion of human cloning, people's attitudes had shifted to accepting animal cloning and gene transfer to humans, whereas germ-line gene therapy had shifted to merely "problematic."[31]

In Chapter 3, I described how James Watson once opposed in vitro fertilization by claiming that prejudicial attitudes of other people would harm children created this way. My theme there was that the prejudice was really in Watson, not in other people, and the very way that Watson was stirring up fear was doing more to create the prejudice than any normal human reaction. Similarly, Leon Kass' recent long essay in *The New Republic*, where he calls human asexual reproduction "repugnant" and a "horror," creates exactly the kind of prejudiced reaction that he predicts.[32] Rather than make a priori, self-fulfilling prophecies, wouldn't it be better to be empirical about such matters? To be more optimistic about the reactions of ordinary parents?

In a more telling argument, Kass claims that a child originated by NST will have "a troubled psychic identity" because he or she will be "utterly" confused about his or her social and kinship ties.[33] At worst, he or she will be like a child of "incest" and may, if originated as a female from the mother, have the same sexual feelings towards the "father" as the mother. An older male might in turn have strong sexual feelings toward a young female with his wife's genome (call this the "Joanna May" syndrome).

If this were so, any husband of any married twin might have an equally troubled "psychic identity" because he might have the same sexual feelings toward the twin as his wife. Yet those in relationship with twins or triplets claim that the individuals are very different and so, hence, are their feelings toward the different twins.

Kass also argues that biological relationships ground social responsibilities, and it is important to keep social taboos on incest and clear lines of responsibility for child support. Given that our society already harms children by its high rate of divorce, extramarital conception, remarriage in blended families, and tolerance of lesbians having children, Kass thinks that human asexual reproduction would be just another part of the overall, worsening mess of the modern child's confused genetic relationships.

Much of that line of criticism simply begs the question and assumes that human NST will be greeted by stigma or create confusion. It is hard to understand why Kass thinks so, once one gets beyond the novelty, because a child created asexually would know exactly who his or her ancestor was. No confusion there.

Besides, this entire argument is hypocritical in our present society. There is overwhelming evidence that divorce hurts children, even teenage children, and no one is making any serious effort to ban divorce. It is always far easier to concentrate on the dramatic, far-off harm than the one close-at-hand. When we are really concerned about minimizing harm to children, we will pass laws requiring all parents wanting to divorce to go through counseling sessions or to wait a year. We will pass a federal law compelling child-support from fathers who flee to other states, and make it impossible to renew a professional license or get paid in a public institution in another state until all child-support is paid. One could also mention a couple who are determined to have their own children, even if each has a 25% risk of cystic fibrosis. Why not ban that, too? After these are done, then we can non-hypocritically talk about how much our society cares about not harming children.

A Point about Control

Much of the worry about unrealistic parental expectations stems from a fear of giving parents too much control over their children and the development of those children. Although this is a legitimate worry, we should see it in a larger context.

Decisions about whether to have children, how many children to have, and when to abort are already being made all the time by parents in our society. True, some parents sometimes make bad decisions, but we accept that risk for the greater good of giving those who care the most the right to make decisions. Besides, what is the realistic alternative? Give such control to legislatures? The courts? Social workers? The latter two are already overwhelmed with custody and child-abuse cases; they certainly can't handle any more cases. And legislatures can only make laws cutting a broad swath; they can't do justice to individual cases.

Another way to put this point is to ask not *whether* to give parents control of asexual reproduction but *to whom* to give such control. It is false to believe that if parents don't have control over NST, nobody does.[34] If human NST is banned, it will be because of the NBAC *Report* and Congress. If there is no "sunset" clause in the ban, control over human asexual reproduction will for a long time pass out of the hands of ordinary people. So control of human NST passes to the federal government and to unelected Commission members. As bioethicist Glenn McGee quotes John Dewey on these reproductive matters: "if intelligent method is lacking, [then] prejudice, the pressure of immediate circumstance, self-interest and class interest, traditional customs, institutions of accidental historical origin, are not lacking, and they tend to take the place of intelligence."[35]

Control over an entire area of human reproduction is an awesome thing to give to anyone. If we don't let ordinary parents have it, we relinquish it to the government.

Increasing Prejudice against the Disabled

If we allow people to perfect their children, society will become intolerant of those with imperfections. If we allow parents to try to attempt to produce perfect babies through cloning, they will reject anything less than a perfect baby. Infanticide and child-abuse will increase. So some have argued.

Again, this is a silly argument. The logic here is the same as claiming that if we allow people to improve their health, society will become intolerant of those who are sick. Why does allowing some people to strive to be better imply that society as a whole will condemn those who have bad luck?

One of the real worries here is when people think several hundred years down the line and they imagine that most children have been born with a great set of genes and live happy, talented lives. Then they think of children who have the rotten luck to not only lack (what today we would consider) normal genes but to actually have a genetic disability such as spina bifida. The worry here is that the genetically-superior, normal people will be contemptuous of those with disabilities.

Again, this argument is way too speculative to be practical. Go back two hundred years and imagine all the objections to innovations that would have seemed reasonable to citizens of 1777: automobiles instead of horses, birth control pills rather than mandatory pregnancy after sex, electricity rather than brute human labor, and antibiotics that doubled and tripled life spans ("What would people do, living so long?" they would've wondered.)

The argument also assumes that in the next centuries we will not discover how to do genetic therapy to cure those genetic diseases. Certainly if we originate smarter people who go into science or medicine, we will discover cures faster.

In the only posted reference this author could find by a person with disabilities, the following message occurred in a debate about the ethics of human cloning:

As a person with spina bifida, I must say it's exciting to think that we know enough about DNA to replicate it now. Further research is needed, though, 'till we get to the point where we can manipulate DNA directly, rather than just transplanting it from vessel to vessel.

The potential harm is far outweighed by the potential good in this case. Well, at least in my humble opinion.[36]

Fostering Sexism

Historically, some feminists feared new developments in assisted reproduction because they believed that such developments would be tools by which men in a sexist society would control women. Although many women appear to really want children and many others appear to want to use these reproductive tools, some feminists argued that these women were really under the influence of husbands and fathers, and will be so in the future with other new, reproductive tools.

Again, this is not a very good argument. For one thing, it takes the view of one political position and equates it to "the" feminist position." Twenty years ago, some so-called feminists criticized in vitro fertilization, fearing that new reproductive technologies were devices developed by men to control women.[37] These feminists feared that women would become Stepford Wives and would choose assisted reproduction only because they were coerced by men in their lives or by the male values of the larger society. Such feminists saw the glorification of pregnancy as condemning women to a traditional biological role.

Such would certainly be the criticism if the first dozen instances of human NST were all male and merely gestated by the wife. The wife would then appear to some to be just the "handmaid" of the husband and male son, to be discarded once the new male is produced. But it might also be true that more than half of the first NST kids would be females, since females have ultimate control over NST gestation.

Whether it be a leftist feminist professor or a crotchety old white male conservative commentator on television, rarely do the learned seem more foolish than when they venture into the waters of personal life to criticize the choices of ordinary people. They forget that Mill's principle does not merely champion an area of personal life free from government interference, but also an area free from moral criticism.

Whether it is an orthodox rabbi, a feminist protestor, or the nosey neighbor, decisions about whom to marry, whom to have sex with, how many children to have, and whether to have children should quintessentially be decisions of private life. Indeed, if these decisions don't define private life, what could? Many arguments by critics here smack of condescension. There is no feeling for the joys of creating and gestating children, or to having a big, extended family.

More generally, wouldn't it be arrogant to judge that women who gestate a male child are always manipulated by their husbands? That judgment itself would be sexist because it suggests that these female gestators can not make free choices.[38] It is easy to beg the question by assuming that choosing to originate children by NST, or by any other form of assisted reproduction, represents coercion, whereas traditional choices do not.

Of course, such critics in their lofty heights usually exempt themselves from the position they inflict on others: if women are generally coerced, then wouldn't the position of female critics of new reproductive technologies also be a product of coercion? Does the nature of a choice itself tell us who is free and who is not?

As said previously, the most dramatic reply to women worried about NST as sexist is to emphasize that NST allows women to originate children without men. Women are much more empowered by NST than men. Women can gestate children with their own genotypes; men cannot.

Finally, and very important, women fearing increased sexism from the introduction of human NST have a knock-down argument to any sexist fantasy about human reproduction that would exploit them: they can simply refuse to get pregnant, refuse to stay pregnant, or refuse to gestate a fetus any more. Unless democracies change into dictatorships, women will retain such rights and hence, most of the expected dangers of new reproductive choices will never materialize. Abortion rights thus emerge as effective deterrents against eugenic madness. Any attempt to do something coercive or silly in human reproduction can be countered by women who simply refuse to carry their conceptions.

Class Injustice

The rich will have better and better kids, who will be smarter, stronger, and more beautiful, while kids of the poor will get dumber, weaker, uglier and carry genetic diseases. The poor and politically powerless have nothing to gain from allowing human NST and perhaps, a lot to lose. So the critic of our class-divided society asserts.

Originating a child by NST could conceivably be a status symbol among the wealthy. It might become equivalent to December family skiing vacations in Colorado or private schools that cost more per year than some of the best colleges.

All that is true, but so what? Class injustice is not changed by preventing the wealthy from buying X rather than Y; it is prevented by not having some people (not to put too fine a point on it) have wealth. If there are wealthy people, and if having wealthy people is bad, then it does not follow that what the wealthy can buy is bad.

If this objection is a good argument against creating humans by nuclear transfer, it is also a good argument against all present forms of assisted reproduction, which are generally (and hopefully!) cash-and-carry operations, as well as against private education and the sale of yachts and private jets. In short, it is better to evaluate the ethics of human asexual reproduction in its own sphere, rather than bringing in the most global and controversial attempts to change the world.

The Slippery Slope

In Chapter 2, I used *Future Shock* to discuss slippery slope arguments that were mere manipulations. Specific slippery slope arguments against human NST have two forms: justification arguments and momentum arguments.[39] The first kind of argument claims that whatever justifies human NST in the original case will also justify human NST in related, but different kinds of cases, and so on, until almost any kind of human NST will be justified. The second kind of argument says that once a line is crossed, where human beings are created from asexual reproduction by copying the genotype of an existing human, then society will be unable to stop crossing more and more lines as it tumbles down the procreative slope. Once we allow a woman to gestate a human embryo originated by NST, we are on our way to a Brave New World. That master of appeals to the slippery slope, Leon Kass, writes in this regard:

> We are urged by proponents of cloning to forget about the science fiction scenarios of laboratory manufacture and multiple-copied clones, and to focus only on the homely cases of infertile couples exercising their reproductive rights. But why, if the single cases are so innocent, should multiplying their performance be so off-putting? (Similarly, why do others object to people making money off this practice, if the practice itself is perfectly acceptable?) When we follow the sound ethical principle of universalizing our choice—"would it be right if everyone cloned a Wilt Chamberlain" (with his consent, of course)? "Would it be right if everyone decided to practice asexual reproduction?" we discover what is wrong with these seemingly innocent cases. The so-called science fiction cases make vivid the meaning of what looks to us, mistakenly, to be benign.[40]

The paragraph above will be used in courses in practical reasoning in which students are challenged to discover the maximal number of mistakes of reasoning.

In roughly reverse order, "everyone" cannot practice asexual reproduction because everyone cannot afford in vitro fertilization and few people would want to do so who otherwise could make babies by having sex (straw man fallacy). Second, and as Kant taught us, what is universalized is not the specifics of the choice (Wilt Chamberlain) but the maxim under which one chooses ("Can I will that all others have the same right as me to choose how to conceive babies?"). There is a very sophisticated discussion in the literature of ethical theory as to what Kant really did teach about universalizability, and it certainly is not the above. Applying Kass' version of Kant's universalization, I could prove that it is wrong for everyone to swim in a swimming pool or to drive a car across a bridge.

Last, there are many human activities for which it does not follow that "If X is permissible, then it is permissible to commercialize X by buying and selling it." Witness human sexuality, donation of blood and of organs, and altruistic surrogacy.

I don't think that fear of a slippery slope is realistic. Its metaphor assumes that once the first change is made, a huge amount of pent-up change will also occur. The most imputed motive for such a pent-up change is the desire of parents to have perfect children.

What defeats this fear is understanding that: (1) only a very few people can afford in vitro fertilization. (2) Of the ones who can and who try repeatedly for a baby, only a few ever take home a baby. For the 24,000 IVF babies created in America between 1978 and 1990, maybe 250,000 couples attempted IVF, roughly 25,000 a year. (3) Of these 25,000 couples, only a few hundred might ever try NST. (4) Of these, how many would ever both (a) want to recreate the same genotype in more than one IVF pregnancy, and (b) convince the physicians of the IVF clinic to allow them to do so? Simply because it is so controversial, most clinics would discourage such an attempt. For all these reasons, there is not going to be any mass movement to originate children by IVF/NST. Hence, no slope of justification will force NST to ever-widening circles of practice.

But here someone can object that the slippery slope is still real, at least as an empirical slope, because once the first cases are allowed, and if it is not illegal to originate by NST, then some people with money will find an IVF clinic somewhere to create a child out of very bad motives, such as narcissism or a desire for perfection in a child.

Well, that is true, and there may be a few cases of this. But nothing now prevents a rich man or rich woman from selecting a mate out of narcissism or a desire for a perfect child, and not only does society not interfere but no one condemns them. So long as the couple assumes responsibility for the child's upbringing, we allow a great deal of liberty to people about creating children. So it is conceivable that a rich man could marry a malleable young wife and convince her to undergo IVF and to gestate, say, over several births, five children of different ages with his genotype. The assumption of the slippery slope is that if something like this occurs, the sky will fall and the fabric of society will be destroyed.

I fail to see how. At worst, they will be curiosities; at best, they will be novelties when they first appear together and afterwards people will grow accustomed to them. As they grow into adults, separate, and go their different ways, few people will ever know they grew up as quintuplets. And the world will go on.

It seems pointless to me to spend more time on such slippery slope objections. They are generally unfalsifiable because the harm they predict is so vague and amorphous. They generally seem to be more attitudinal reactions cloaked in the language of an empirical claim than real predictions.

Nature versus Nurture

A final objection that I will discuss and refute is a bit strange. The political argument is sometimes heard that human NST would be wrong because it will lead humans to place too much emphasis on genetics ("nature") and not enough on changing the environment to reduce the natural inequities of fate ("nurture"). People who are progressive liberals sometimes say this kind of thing.

Such a liberal seems to be a genetic reductionist because what he really is thinking is that genes are almost everything, but we shouldn't admit it. Instead, we should keep trying to pretend to change what is really unchangeable. On the other hand, if we admit the truth, it will depress those who have lost out in the genetic lottery.

Such genetic reductionism is becoming common in popular thinking, especially as people begin to think that being gay or lesbian is just a matter of having a specific gene, or that all disease is in the genes.

In reply, those seeking more genetic knowledge believe that human control will ultimately create better humans. It is fascinating to think that origination by NST of even one of a few children from the same ancestor might answer some questions in the nature/nurture debate.

Of course, people should never be created simply to be the subjects of scientific experiments. People originated by NST, as I have emphasized, should have the same moral and legal rights as any other kind of people. But once they existed and were raised in ordinary families, facts about the lives of people originated by NST would provide new answers to questions in this ancient debate.

At present, a huge amount of money is invested on the assumption that fetal environment, early childhood learning, nutrition, and better schooling affect how children turn out. That assumption could turn out to be false. It may be that the money and effort are all wasted. Although that prospect may depress some people, it is better to have a national policy based on truth than on ignorance.

Simplistic responses are tempting to make, but usually incorrect. In advance, we should be careful about assumptions that any trait or ability is genetic or environmental because our assumption might be used against us. Historian Robert N. Proctor notes that cigarette companies once argued that whether a person got cancer was genetic and had nothing to do with cigarette smoking.[41]

The final truth may cut either way. It may be that such programs as Head Start, prenatal care, alcohol-awareness for pregnancy mothers, and an emphasis on talking to kids in the first year of their life are *much more* important than we previously thought. Now we are groping in darkness, and in such a situation, any light will allow us to move forward some, even if what we see is not what we expected.

Knowledge can also open up new options. Currently, most people believe that individuals should control their eating to control their weight. This may be very hard for some people. Currently, some individuals use medical techniques such as liposuction or surgical shrinkage of the stomach to alter their appetite and bodies. If this is acceptable, would it be acceptable to introduce a gene into one's own body to change one's metabolism? If it were possible to get lots of information about the side-effects of such genetic insertion and if it were possible to fully deliberate about such a choice, wouldn't allowing humans to so choose allow them to maximally express their autonomy?[42]

In short, it is hard to take the argument seriously that everyone is better off in ignorance. Truth has a way of yielding unexpected results that do not fit neatly into preconceived social-political theories.

Conclusion

There are many arguments against human NST. It is claimed that this new form of human creation would be against divine will, would cross a line into unknown dangerous territory, and would be of possible harm to the child either from developmental defects or unrealistic expectations.

I have argued that none of these arguments work. The first kind of argument revolves around fatalism and the idea that it is dangerous for humans to make new kinds of choices. This argument has ancient roots but all of modern medicine refutes and opposes it.

The second kind of argument revolves around various possible harms to a child originated by NST. Again, almost all such fears are more imagined than empirically likely. Should a decade of mammalian studies on babies produced asexually show no greater rate of birth defects, we will be ripe to try to originate a NST baby.

By far the most widely-believed objection is that such asexual reproduction will damage the child in some social way. Here it is important to emphasize the similarly incorrect claims made twenty years ago about in vitro fertilization.

The rest of the arguments, which rely on concerns about sexism, the disabled, and class injustice, are no better, begging too many questions to be relevant. Finally, slippery slope arguments in bioethics are usually just clubs that people use who don't want to change, and their use in opposing human NST is no different.

Notes

1. National Bioethics Advisory Commission (NBAC), *Cloning Human Beings: Report and Recommendations of the National Bioethics Advisory Commission*, Rockville, Md., June 1997, 56.
2. NBAC, *Cloning Human Beings*, 56.
3. NBAC, *Cloning Human Beings*, 59.
4. Quoted in the NBAC *Cloning Human Beings*, 47, from Paul Ramsey, *Fabricated Man: The Ethics of Genetic Control* (New Haven, Conn.: Yale University Press, 1997), 136.
5. Gilbert Meilaender, "Begetting and Cloning," *First Things* 74 (June/July 1997), 41–43.
6. NBAC, *Cloning Human Beings*, 43–49.
7. George Annas, "Scientific Discoveries and Cloning: Challenges for Public Policy," Senate Subcommittee on Public Health and Safety, Committee on Labor and Human Resources, 12 March 1997. Testimony posted on the Internet at this address: http://www-busph.bu.edu/Depts/LW/Clonetest.htm.
8. Lori Andrews, moderator, "Legal, Regulatory, and Industry Issues," Conference on Mammalian Cloning: Implications for Science and Society, 27 June 1997, Crystal City Marriott, Crystal City, Virginia.
9. See Leroy Walters and Julie Gage Palmer, *The Ethics of Human Gene Therapy* (New York: Oxford University Press, 1997), 36ff.

10. Robert Lee Hotz and Thomas Maugh, II, "Biotech: The Revolution Is Already Underway," *Los Angeles Times*, 27 April 1997 (Internet edition).

11. Quoted in Shankar Vedantam, "After Initial Opposition, Some Now Defend Human Cloning," Knight-Ridder newspapers, 10 March 1997.

12. Diana Lutz, "Hello, Hello, Hello, Dolly," *The Sciences*, May/June 1997,10.

13. Paul Ramsey, *Fabricated Man: The Ethics of Genetic Control* (New Haven, Conn.: Yale University Press, 1970).

14. Richard McCormick, quoted by Gustav Niebuhr, "Cloned Sheep Stirs Debate on Its Use in Humans," *New York Times*, 1 March 1997 (Internet edition).

15. C. S. Lewis, *The Abolition of Man* (1943, many editions). This is the book in which Lewis defends his version of natural law.

16. Joseph Fletcher, "Ethical Aspects of Genetic Control," *New England Journal of Medicine* 285, no. 14 (1971), 776–777.

17. Jonathan Glover, *What Sort of People Should There Be?* (New York: Penguin, 1984).

18. Leroy Walters, "Biomedical Ethics and Their Role in Mammalian Cloning," Conference on Mammalian Cloning: Implications for Science and Society, 27 June 1997, Crystal City Marriott, Crystal City, Virginia.

19. A. Wilcox et al., "Incidence of Early Loss of Pregnancy," *New England Journal of Medicine*, 319, no. 4 (28 July 1988), 189–194. See also J. Grudzinskas and A. Nysenbaum, "Failure of Human Pregnancy after Implantation," *Annals of New York Academy of Sciences* 442 (1985), 39–44; J. Muller et al., "Fetal Loss after Implantation," *Lancet* 2 (1980), 554–556.

20. See the web site of a leading law firm in this area, Elman & Associates at http://www.elman.com/elman.

21. Joshua Lederberg, "Experimental Genetics and Human Evolution," *American Naturalist* 100, no. 915 (September-October, 1966), 527.

22. Harold Varmus, *New York Times*, 13 March 1997.

23. Mary Midgley, *Evolution as a Religion: Strange Hopes and Stranger Fears* (London: Methuen, 1985).

24. NBAC, *Cloning Human Beings*, 22–23. This list is also based on an overview talk by molecular biologist Lee Silver, "The Current Status of the Science," Conference on Mammalian Cloning: Implications for Science and Society, 27 June 1997, Crystal City Marriott, Crystal City, Virginia.

25. A. Wilcox et al., "Incidence of . . ."

26. One might object, "If we cannot *harm* a child that doesn't exist, can we *benefit* a child that doesn't exist?" It seems to me the answer is, "Well, yes, you can be benefitted by being made to exist. Consider the alternative."

27. Axel Kahn, "Clone Animals . . . Clone Man?" specially-commissioned article to accompany articles in *Nature* on cloning on the web site of *Nature*.

28. Alex Capron, "Legal and Social Issues and Review of the NBAC Report," Conference on Mammalian Cloning: Implications for Science and Society, 26 June 1997, Crystal City Marriott, Crystal City, Virginia.

29. Peter Singer and Deane Wells, *The Reproductive Revolution: New Ways of Making Babies* (New York: Oxford University Press, 1984), 161.

30. Leon Kass, "The Wisdom of Repugnance," *The New Republic*, 2 June 1997, 22.

31. Ron James, Managing Director, PPL Therapuetics, "Industry Perspective:

The Promise and Practical Applications," Conference on Mammalian Cloning: Implications for Science and Society, 27 June 1997, Crystal City Marriott, Crystal City, Virginia.

32. Leon Kass, "The Wisdom of Repugnance," *The New Republic*, June 2, 1997.

33. Leon Kass, "The Wisdom. . . .", 22–23.

34. I'm indebted to Jim Rachels for this point.

35. Glenn McGee, *The Perfect Baby: A Pragmatic Approach to Genetics*. (Lanham, Md.: Rowman & Littlefield, 1997), 132. John Dewey quotation from John McDermott, *The Philosophy of John Dewey* (Chicago, Ill., University of Chicago Press, 1981), 583.

36. Message #103, "Send in the Clones," *New York Times* Forum on cloning on its Internet server.

37. Mary Ann Warren, "In Vitro Fertilization and Women's Interests: An Analysis of Feminist Concerns," *Bioethics* 2, no. 1 (January 1988), p. 44.

38. Christine Sistare, "Reproductive Freedom and Women's Freedom: Surrogacy and Autonomy," *Philosophical Forum: A Quarterly* 19, no. 4 (Summer 1988).

39. Douglas Walton, *Slippery Slope Arguments* (New York: Oxford University Press), 1992.

40. Leon Kass, "The Wisdom of Repugnance," 25.

41. Robert N. Proctor, "Resisting Reductionism from the Human Genome Project," in *Classic Works in Medical Ethics: Core Philosophical Readings*, ed. Gregory E. Pence (New York: McGraw-Hill, 1997), 347ff.

42. On this point, see P. S. Greenspan, "Free Will and the Genome Project," *Philosophy & Public Affairs* 22, no. 4 (Winter 1993).

Regulating Human Cloning

Human cloning will be driven by the market. Everybody is suggesting it will come through the private clinics, which are not regulated and are used to giving the customer what he or she wants, operating in many states on a philosophy of "Just show me the money."

Lori Andrews, Professor of Law, Chicago-Kent College of Law[1]

· · · · ·

There are some areas of the law that are questions for lawyers' lawyers: finely-grained questions about contract and commercial torts, based on nuances built up over hundreds of years and thousands of cases. In such an area, everybody knows the parameters, e.g., whether victims of car accidents can sue and how, and where morality enters, it entered long ago in setting up those parameters. Inside them, the questions are mainly legal, although moral questions can still arise if someone crosses a line, e.g., ambulance-chasing lawyers.

On the other hand, there are questions where the boundaries are being debated. It is here that philosophical analysis and ethical debate shapes things. Thinking about originating a person by NST is one such area of the law, the subject of this chapter.

In this chapter, I first review arguments that we should not regulate human asexual reproduction in any form. I next review the arguments for regulation, a position I ultimately accept, although with qualifications. I also argue in this chapter that we should ban the buying and selling of human genotypes.

The John Moore Case

In 1988, John Moore lost a suit in the California Supreme Court. Moore was a Seattle native who had his leukemia treated at a California medical center. A physician, David Godly, patented a cell line created from tissue of Moore that was thought to be valuable in creating treatments against leukemia.

Moore brought suit to share in the profits from a product of his body. The California Supreme Court ruled against him, arguing that people like Moore have no property right in their cellular material. The decision stated that Moore's informed consent was required before such material could be taken from him.

This decision, although technically binding only in California, set a precedent and has been widely cited. It might indicate that courts are likely to hold that one cannot buy or sell one's own genotype for creating embryos by NST, but that if such a procedure is done non-commercially, then the originator of the genome must give informed consent. Minimally, this would mean that the person who is the source of the genome was informed of the purpose of using his cells to create another person with his genes.

The Case against Regulation of NST

Libertarians argue that government regulation of any area of personal reproduction is a disvalue that must be balanced by some greater good. It is an evil first because it is an exercise of control over human procreative liberty and second because it thwarts the ability of the market to allocate resources where humans actually want them to be (as opposed to where some anonymous bureaucrat says they should be). So regulation of human reproduction is a disvalue because it may severely restrict human goods.

Of course, we are talking of regulating reproductive services here, not goods, but even in this case, regulation can be a bad thing. Take the example of in vitro fertilization. One year after the birth of IVF baby Louise Brown in 1978, an American Ethics Advisory Board recommended that the government fund studies to prove the safety of in vitro fertilization. Presumably, until IVF could be proved safe, a moratorium would be imposed. The report has gathered dust for nearly 20 years and never had any effect. Why? Because the birth of American baby Elizabeth Carr in 1981 at the Jones Institute in Norfolk, Virginia, put an end to worries about safety. If a picture is worth a thousand words, a healthy baby is worth a million.

If IVF clinics come under federal regulation, such clinics and their physicians will likely become the gate-keepers of federal control over any new kind of human reproduction, including human NST. As mentioned previously, originating a human by NST can only be done with the assistance of specialists at such a clinic. Getting the NST embryo formed is only part of the problem; the other part is getting it to implant, stay implanted, and gestate without miscarriage

Some infertility clinics at present already control access to their services. They impose various conditions on people seeking their assistance: most have an age limit for the mother, will not implant embryos of only one sex, some will not (rightly or wrongly) implant embryos in women who are unmarried or part of a lesbian couple, and so on.

But it is not apparent to some why any of these restrictions are justified. There is no other area of medical services where physicians or the public feel they have the right to restrict medical services offered to other people. Second, if human reproduction is a good, why is it only a good to people of a certain kind? Third, many of the restrictions are not based on evidence but the prejudices of the day.

There is something very scary in some of the stuff being advocated by some law professors. George Annas has proposed a "Human Experimentation Agency" that would put the burden of proof on any physician wanting to do a new kind of experiment.[2] This new federal bureaucracy would pass on which experiments were allowed and which were not.

Federal bureaucracies are fine in some areas, such as protecting the environment, but when they get into sex, abortion, and the bedroom, they are way too politicized. Like NIH, they are too susceptible to pressures from extreme religious groups.

Annas and other lawyers are comfortable with adversarial processes and believe that only if there is an adversarial process can rights be safeguarded. So he doesn't think physicians and scientists can be trusted to follow professional ethics and professional guidelines: "We cannot expect physician-dominated groups to protect the interests of patients any more than a guard-dominated group would protect the interests of prisoners, a landlord-dominated group would protect the interests of tenants, or a police-dominated group would protect the interest of citizens."[3] But is it fair to lump scientists and physicians together with prison guards? Inmates and prison guards have contradictory interests; reproductive physicians and parents getting together to create a child by NST do not.

As for experiments on human NST, Annas' new Human Experimentation Agency would work this way: "cloning proponents should have the burden of proving that there is *an important societal purpose* for such an experiment, rather than the regulators having the burden of proving that there is some compelling reason not to approve it"[4] (emphasis added).

I don't know about the reader, but the italicized phrase scares me. How would a couple and an investigator prove to an august federal agency that their desire to originate a child by NST served an "important societal purpose"? When have Americans ever been subjected to such a requirement before they were allowed to have children? In discussing arguments against human NST, I previously argued that someone always controls human reproduction and that to ban human NST gives all the control to the government. Annas' statement makes it very clear what this new federal agency would require of citizens and investigators. Something like this was once seriously proposed about how abortions should be legalized. A woman would go to a clinic and be required to explain her reasons for wanting an abortion. If her reasons were good enough, she would be allowed to abort. Otherwise, she would not.

We know enough about the politicization of abortion to know that this would have been a bad policy. One woman's good reasons are another man's insufficient reasons. We can imagine that clinics run in such a way might have been required, in contested cases, to hear the male's reasons for opposing the abortion, and to do so, the male would have to be notified and have the chance to appear. And so on. What a nightmare!

It is worth noting that European bioethics tends to emphasize central planning and legislation, which are clumsy all-or-nothing tools for shaping human actions. Scientists and consumers in market-based democracies tend to loathe such central planning, be it legislation or regulation, and prefer that the preferences of consumers determine what is done. In areas of gene therapy and reproductive ethics, European countries have attempted to anticipate all future contingencies in their design of laws and regulations. America has been more empirical and case-based, with more attendant flexibility. There are advantages and disadvantages of each approach. As one law professor says, "The European system tries to design the dog and let it wag its tail. We (in America) have 50 or 100 or 150 wagging tails from which we then try to reconstruct our dog."[5]

Regulation certainly has its disadvantages, as the above arguments make clear. Is there any way to gain the advantages of regulation without the disadvantages?

The Case for Regulation

I believe it would be a mistake to pass laws banning nuclear somatic transfer. Because these issues are linked in the minds of some to abortion and to attacks on motherhood, they inflame too many passions. As such, once human reproduction by NST is banned, it will be very difficult to later undo such a ban.

NBAC wants such legislation, but with a sunset clause. The history of past attempts to get such a qualification is not impressive. The 1994 Human Embryo Panel wanted Congress to accept some embryo research in carefully-defined areas. What it got was a blanket rejection of all such research. Some issues defy legislation with careful distinctions, and originating humans by NST is one such area.

On the other hand, I think regulation done in a limited way might work and would be justified. The best way would not be a new federal agency but bringing infertility clinics under the loose supervision of the Centers for Disease Control. After some conferences to get input from all sides, guidelines could be made for such clinics and revised every three years.

Robert Cook Deegan, Director of the National Policy Board for the Institute of Medicine of the National Academy of Science, argues that we should model the control of human asexual reproduction on the model of gene-therapy, which has successfully supervised genetic therapy over the last decade by regulation without legislation.[6] I think he is correct. Initially feared, somatic genetic therapy is on its way to becoming an accepted part of medicine.

Through regulation of IVF clinics, two important questions can be discussed and answered: (1) When is it safe to originate a child by NST? and (2) Will it work when we try to do so? Everyone can have their say about this in public discussions, including scientists and representatives of religions.

There is a great historical model for how such regulation can be successful. The regulation of recombinant DNA research, the big controversy of the late 1970s, was contained by voluntary efforts of leading scientists in convening the famous Asimolar conference. Such research now is only supervised by local Institutional Review Boards (IRBs) and the NIH Recombinant DNA Advisory Committee (RAC), which advises NIH on the regulation of gene therapy. The RAC has functioned as a national ethics committee for a particular modality of treatment, viz., for human somatic gene therapy. Under this plan, no abuses have occurred and the public opinion now accepts somatic, gene therapy.

To be fair, there are those who see things differently. Some researchers believe that RAC review has been way too cautious and too slow.[7] People now see somatic gene therapy as just another kind of medicine. Hundreds of cancer protocols go forward with much less review. Isn't it time, critics ask, to loosen up RAC review of somatic gene therapy? Especially where the Federal Drug Administration must also review their protocols, subjecting such researchers to tedious dual review. In general, regulation may be good in keeping controversy to a minimum, but it also has its price in slowing down daring attempts to find cures. If every great scientist in the past had to pass his proposals through two committees of peers, would we still have all our victories over disease?

Germany decided to legally ban, not regulate, and we can see the pitfalls in that approach. Germany passed legislation that shifted the onus of proof to scientists and physicians. The default condition in Germany is to ban the procedures and use them only if evidence can be mounted proving them safe. It is a federal crime there to attempt any kind of genetic therapy.

Along this line, it is important to note that the European version of NBAC, meeting in The Hague about the same time as the NBAC *Report* appeared, rejected even embryo twinning to increase pregnancy rates for in vitro fertilization. The committee decided that European women should be subjected to taking powerful, possibly cancerous drugs to stimulate superovulation rather than take the "risks" of twinning embryos. To allow such twinning, it said, would risk a slippery slope.[8]

When the German law was passed, no conceivable therapeutic case could be imagined for genetic therapy. But knowledge accumulates fast, especially with the Human Genome Project, and now we know that there are some genetic diseases related to mitochondrial DNA that could be reduced by NST, especially from an embryo produced sexually and from which a nucleus is then transferred to an enucleated egg. To help German couples and children with these problems, a federal law must be overturned—not an easy undertaking. Far better not to build this obstacle into medicine and scientific research in the first place.

Leroy Walters argues that regulating NST rather than legislatively banning it could have many beneficial effects.[9] Again, I agree. For NST research, regulation would impose a single set of standards on all American clinics. Regulation would require some supervision of each NST experiment. Regulation could be supervised by a national oversight panel, such as the NIH peer-review panels that now judge applications for NIH funding. Regulation would allow public supervision, education, and control of publicly-funded NST research. Ideally, such regulation would have a sunset provision, allowing it to expire after a half-dozen years.

For the above reasons and others, I favor regulation of NST in infertility clinics, but only of those IVF clinics that want to do NST, for the first years of applying this technique. In addition to the above reasons, I think that humans are complex enough that we should go slowly to understand all the things that can go wrong.

For example, troublesome cases that have already occurred about embryos and are worth reviewing for what they augur about persons originated by cloning. The gee-whiz issues might be off-base but the perennial issues of bitter divorces and disputes about custody of children are certainly bound to arise.

In 1981, the issue of embryos was dramatized by a famous case, that of Mario and Elsa Rios. The Rioses, an American couple worth millions, traveled to Melbourne, Australia, where several of Elsa Rios's eggs were fertilized in vitro with sperm from an anonymous donor. Three embryos were produced; one was implanted, and the other two were frozen in a tank of liquid nitrogen for later implantation in case the first attempt failed. Ten days later, the first attempt did fail, but the Rioses were unwilling to make a second attempt immediately and returned to the United States. They later went to South America. In the spring of 1983, they were killed in a plane crash. The story made headlines when it was learned that their frozen embryos still existed. Neither the Rioses nor the infertility clinic had provided for this contingency, and the legal status of the embryos was doubtful.

Several ethical questions arose. Could the embryos simply be destroyed? If they were implanted in surrogate mothers and carried to term, could they later sue in American courts for their inheritance from the Rioses? Should the anonymous sperm donor be consulted about his wishes? An ad hoc committee of the Australian government recommended that the Rios embryos be destroyed, having concluded that removing their life support would be the same as removing life support of terminal patients. The Australian parliament, however, reasoned differently and required that the embryos be preserved until each of them could be "adopted." Presumably, they have since died (because they were frozen at probably the wrong temperature for long-term safety).

In a highly publicized case in the early 1990s, Mary Sue Davis and Junior Davis of Tennessee divorced and fought for custody of seven embryos frozen in their IVF clinic. Initially, Mary Sue Davis sought custody so that she could use the embryos to become a mother, whereas Junior Davis wanted custody to prevent her from becoming pregnant with embryos with his gametes. A lower court judge awarded custody to Mary Sue Davis; but in June 1992, the Tennessee Supreme Court decided that Junior Davis did not have to become a father against his will. Junior Davis won in part because both parties had remarried and after her remarriage, Mary Sue Davis wanted the embryos in order to donate them to some other infertile couple. In 1993, the United States Supreme Court declined to hear an appeal, but a Tennessee court later ruled, in another hearing, that Junior could destroy the embryos, which he did.[10]

These cases provide one with grounds for making certain predictions about humans originated by NST. Children originated by NST will probably grow up in marriages that end in divorce at the normal rate for American couples. Some of these marriages will have disputes about custody of children. The twist for children from nuclear transfer is whether the genetic ancestor of the child will be held to have more of a claim than the other parent. If the child was a female identical in genes to her mother and gestated by her mother, it would seem that courts would have a difficult time giving custody to the father. On the other hand, if the child were male, the father's bid might trump the mother's because of the unique genetic connection.

The Issue of Multiples

So far in this book, I have tried to separate nuclear somatic transfer from multiple copies of a genome, each created by NST or NST plus twinning. I think there are good reasons for doing so, both conceptual and prudential. Conceptually, we can separate the intrinsic objections to NST by setting aside different issues about multiple originations of the same genome. Prudentially, it may be wise to limit multiples until the dust settles, for reasons that I now explain.

First, the scariest public fears about cloning involve multiples. Such originations raise all the old fears about "armies of clones," manufacturing humans, and losing human specialness. Although I do not object to multiples on principle, I think that we can distinguish between principled and prudential opposition to multiples, and my objection stems from the latter.

Pragmatically, infertility clinics must limit the number of embryos allowed to gestate in any one uterus. Multiple-embryos must sometimes be "selectively reduced" to guarantee birth of fewer, healthy babies. With too many embryos inside a uterus, nutrients become a scarce resource and, suddenly, we have too many embryos in the "lifeboat" where not all can live. For people who persist in trying to gestate large numbers of embryos, birth deformities in resulting babies are often the result. Finally, one would not want to try to gestate a large number of the same genome in one gestation because the varied genes and characteristics of different embryos are precisely what allows some to survive when others do not. If multiple copies of one genome were implanted, it would tend to create an all-or-nothing situation.

Third, there are moral and legal problems about multiples that have yet to be worked out. Does one have the right to control reproduction of one's unique genome? It would certainly seem so as part of one's general personal rights, and it would seem wrong to reproduce someone's genome (e.g, from a blood sample) without his explicit, informed consent. But how can such a right be protected? Once an embryo is created from the genome of X, it can be twinned and re-twinned. Am I wronged if I consent to one child being created from my genome and then ten are created?

Finally, perhaps the expectations argument should have this much weight: maybe we should originate children individually by NST for a few years until they are accepted, and then allow multiples. In this way, the public can adjust to NST children, and these children can adjust to their new status without the oddity of seeming so different.

Ultimately, I don't think the problem of multiples is great. As emphasized previously, because of mitochondrial DNA, different uterine environments, and different "total environments" as soon as each baby is born, we are not really going to have several babies that are identical to each. Once the genomes are gestated by different women in different years, small differences at early ages will become big differences later (as probably happens now with conjoined twins). But it will take society a while to learn all this, and for prudential reasons, it seems best to go at first slowly. One task of regulation would be to ban multiples in infertility clinics for the first few years.

Against Commercialization of NST

I also favor banning the sale and purchase of genomes for origination by NST. It is way too early to understand what it might mean to have a market in genotypes.

If we had decades of success in buying and selling organs for transplantation, we might now have an appropriate experience to think about a system for selling genomes. But we don't. Moreover, we do have some experience with buying and selling blood, and one thing we know from that experience is that people desperate to sell their blood are not always motivated to disclose the truth about whether their blood might be contaminated by HIV, hepatitis, or other conditions that would make their blood unsalable. Similarly, if a person could sell his genome, would he or she disclose that brush with suicide twenty years ago? Would it be fraud not to disclose it?

To buy and sell a genome implies that a product is being sold. A product can be delivered without defects or in very defective shape. Suits would occur when "products" were born with defects. Our system of law and ethics has only begun to think all this out. For now, it seems to me that the primary focus should be on the best interests of any child created by NST, and I do not think those interests are served by commercialization. Without commercialization, people must find altruistic "donors" of genomes (who, undoubtedly, will usually be one of the parents who will raise the child, but not always). These donors are much more likely to be properly motivated if they receive no money for their donation.

We have another recent experience with buying and selling something similar. Using laparoscopy, surgeons can remove many ripe eggs from ovaries of young donor women for implantation in older women.[11] *Sixty Minutes* revealed in 1996 that some young women are paid as much as $8,000 for their eggs.[12] An "egg broker" matches prospective sellers to prospective buyers. Lack of success of IVF in women over 40 who have failed to conceive after several attempts has created this market. Using the eggs of young women with sperm of their husbands, gives older women higher rates of success. (The 68-year-old woman who carried a fetus to birth in 1997 used such an egg from a younger woman.)

I do not think that this has been a good practice. Couples seeking a young egg donor focus too much on how much the young woman looks like the older woman, and that is often to miss the forest for the trees. This is exactly the reason that William and Elizabeth Stern chose Mary Beth Whitehead as their surrogate mother, ignoring a psychological profile that said Mrs. Whitehead would have difficulty giving up her child.[13] Some experience would be good before society allowed couples to do this, so that people could be educated to learn that, say, buying a genome from Michael Jordan does not necessarily mean that you get a person who will be an NBA superstar.

For another thing, and again, as soon as someone sold his genome to create an embryo, it is difficult to see how his property rights could continue to be enforced, because the owner of the new embryo could twin it, and re-twin it, or take one-cell for sale to create another NST child. For that matter, how do you prevent people from using your blood to create a child by NST? As the 1997 movie *GATTACA* made clear, it is hard to shield your DNA from those who want to get it. There are questions within questions within questions for the law to consider (can you patent your genome? trademark it?), and we are not nearly ready to answer most of them, so commercialization should be banned until we know how to protect any children whose genomes have been bought.

Conclusion

Regulation of human reproduction by NST is justified in an experimental phase where protocols are conducted to determine how to make it safe for subsequent children, to limit multiples, to allow public understanding to grow, and to have some national standards of good research. Once a record of safety has been achieved, and once the public accepts NST babies, regulation is no longer justified and human NST should be left to private clinics.

Reluctantly, I accept that this means that some IVF clinics must now fall under regulation. However, my guess is that very few such clinics would want to try NST, and as such, most could continue to be unregulated.

Notes

1. Katherine Q. Seelye, "Clinton Bans Federal Money for Efforts to Clone Humans," *New York Times*, 5 March 1997 (Internet edition).

2. George Annas, "Regulatory Models for Human Embryo Cloning: The Free Market, Professional Guidelines, and Government Restrictions," *Kennedy Institute of Ethics Journal* 4, no. 3 (September 1994), 235–249.

3. George Annas, "Regulatory Models . . . ," 243.

4. George Annas, "Regulatory Models . . . ," 246.

5. Alto Charo, quoted by Sheryl Stolberg, "Reproductive Policy," *Los Angeles Times*, 29 April 1997 (Internet edition).

6. Robert Cook Deegan, "Legal, Regulatory, and Industry Issues," Conference on Mammalian Cloning: Implications for Science and Society, 27 June 1997, Crystal City Marriott, Crystal City, Virginia.

7. See Leroy Walters (with Julia Gage Palmer), *The Ethics of Human Gene Therapy* (New York: Oxford University Press, 1997).

8. Robin Herman, "European Bioethics Panel Denounces Human Cloning," *The Washington Post*, 10 June 1997 (Internet edition).

9. Leroy Walters, "Biomedical Ethics and Their Role in Mammalian Cloning," Conference on Mammalian Cloning: Implications for Science and Society, 27 June 1997, Crystal City Marriott, Crystal City, Virginia. See his (with Julia Gage Palmer) *The Ethics of Human Gene Therapy* (New York: Oxford University Press, 1997).

10. Associated Press, "Ex-Husband Has Embryos Destroyed," 16 June 1993.

11. Sheila Anne Feeney, "Overcoming Infertility," *New York Daily News*, 31 December 1990, D2.

12. "Egg Brokers," *60 Minutes*, 1996.

13. See "Surrogacy: Baby M," in Gregory Pence (ed.), *Classic Cases in Medical Ethics*, 2nd ed. (New York: McGraw-Hill, 1995), 123.

Conclusions

· · · · ·

If the reader is not sympathetic by now to the idea of permitting the origination of a human child by nuclear somatic transfer, there is little more that I can do with rational arguments. In this last chapter, I expand a few themes of this book, and then I turn to something different. In the very final sections, I try to imagine some alternative pasts and futures to give us new intuitions that are more sympathetic to human asexual reproduction.

The Quality of the Arguments in the NST Debate

This effort to counter arguments against NST seems endless, because as soon as one objection is countered, opponents concede that it has been countered but shift to a new objection. This phenomenon resembles an observation made by the libertarian Harvard philosopher Robert Nozick, who notes that critics of capitalism will often concede a specific rebuttal to a specific objection to capitalism, but then immediately move to another objection.[1] Even after many such rebuttals and shiftings have occurred, the critics are not ready to accept capitalism. Nozick wryly observes that perhaps something else motivates the rejection of capitalism than the truth of the arguments.

In this regard, it is frustrating to deal with the arguments in the NBAC *Report.* Some of them contradict each other. Most NBAC members knew that NST to create a human baby would only be used by a few people wealthy enough to afford both in vitro fertilization attempts and nuclear transfer technology, yet they granted full weight to objections based on "predictions of the widespread [negative] effects on society should this type of cloning become a frequent practice."[2] Similarly, a person who opposes NST because it will affect the human gene pool may contradictorily object to it also because it will be used only by the wealthy.

In a similarly frustrating vein, the NBAC *Report* emphasized that genetic determinism was false ("Indeed, the great lesson of modern molecular genetics is the profound complexity of both gene-gene interactions and gene-environment interactions. . . ."[3]) So in the section "The Science and Application of Cloning," the *Report* expresses the view that (I) a child originated by NST would be unique. My problem with the NBAC *Report*, as well as the speakers at the cloning conference following its publication, is that everyone then went on to condemn human NST because "most people" would treat a person originated by NST as a copy with a predetermined future.

In other words, the Commission and the speakers held that a good argument against human NST was that most people would reject the truth of (I) above. Now that is a very strange way to make public policy. Generally, we do not think it a good idea to make law on falsehoods. Here again is where reasons must come into ethics and we must make those reasons trump our immediate emotions.

Simply put, it is self-defeating to say out of one side of your mouth that X is false and then argue that because so many people believe X, all our laws and public policy should be based on X. Obviously, the rational thing to do is to educate people that X is false, not base policy on it.

The Unreality of the Human Embryo Debate

Perhaps you are different from me and, before human cloning came along, you paid a lot of attention to the debate in medical research and bioethics over what can and cannot be done with human embryos. But I doubt it.

To me, it is a fact that only a tiny number of people really believe that such embryos are persons with a right to life. I further believe that most of those who say they do are misinformed. People who worry about the welfare of human embryos falsely imagine that the embryos are really tiny humans, like homunculi.[4]

If these champions of embryos are savvy enough to know that embryos less than a hundred cells don't show human shape, they may believe the linkage argument that if we allow destruction or research on such embryos, we will do the same on fetuses. The very fact that we allow abortions but no federal research on embryos shows that the linkage argument is false. However, I don't want to get into the morality of abortion here and its many issues.

What I do want to ask is this: how are we ever going to move past the ban on studying human embryos? With what Wilmut has discovered, there are many great avenues of research now open to science that, for the best results, would involve the use of human embryos. NIH can't fund such research, and that means that 95% of the money for such research is revoked. Which means not much will get done. Add to this argument that we could help infertile couples have kids by increasing the supply of embryos through twinning, and it's hard to see why the debate has not moved beyond this point.

In bioethics, the view of this tiny minority gets too much attention and is given too much ethical weight. When presenting a talk on ethical issues about abortion, each ethicist must admit that beliefs about such embryos are "an issue to some." But then that issue gets left hanging there, and not countered, so it gathers momentum. Then no one wants to come out against such views, because the point man here must always deal with the zealots, a very unpleasant task. (Politicians know this too well, hence their usual cowardly stands and courtship of the embryo-saving crowd).

Somehow, some way, we must get past the point of worrying about the rights and good of a few thousand human embryos. (There are already thousands in existence, frozen in limbo in liquid nitrogen and no great evil will occur if there is a massive thaw.) Perhaps one good that will come out of the debate about human nuclear somatic transfer is that it will defuse the issue of research on human embryos. Compared to what most people (incorrectly) believe about human cloning, research on human embryos should seem innocuous.

What Might a Good Religious Objection to NST Look Like?

To date, almost all objections to NST from a religious viewpoint have been based on fear of some kind: fear of change, fear of changing human nature, fear of humans having more choice and control over their reproduction, and so on. Obviously, progress cannot occur based on fear, but is there any kind of objection that a religious person might make that is justified?

To me, the most likely, justified objection of this kind would come from the agapic traditions in Christianity and Judaism emphasizing not fear and irrationality but love and reason. From this orientation, people would ask foremost: what is best for the child? Of importance, but much less importance, people should ask: what is best for the family? (As for what is best for "society" or "the community," these questions imply too much control over the family or creation of the child to be good questions. Besides, if it's good for the child and good for the family, society will be fine.)

Along this line of reasoning, it seems to me that the most powerful, religiously-based argument against NST would be to give reasons why NST was bad for the child or bad for the family of a child created by NST. For example, if an NST didn't have a soul or was retarded, then this would be a good argument against allowing NST (but of course, we have no evidence for those claims). On the other hand, if creation of children by NST led to the practice and attitude among sophisticated families that they had some control over the kinds of children they could create, and these families rejected the view that whatever child was born by random genetic mixing should be accepted fatalistically, I do not think that this would be a good religiously-based objection. The agapic tradition in Christianity is not hostile to giving families more control over procreation and, so far, we have not heard much from this tradition about NST.

To put this point differently, that part of Western religions based on fear and control always approaches human reproduction with "Thou Shall Not!" What is needed is a vision of ideal reproduction that embraces new ways of making babies and that says, "Thou May!." It is fruitless for clergy to always be saying "No" without giving alternatives. It's so much better to say, "Not this way but that way." However, to do so, there must be some vision of how originating a child by NST would be acceptable and proper.

To continue to say "No" to any form of NST is to leave it to humanists and scientists to figure out the ethics of NST. This leaves religion with no real voice to play. It is like what has happened to the Vatican, with its preposterous, continuing claim that a wife and husband who deliberately originate a child by IVF are guilty of a sin: people who hear this view expressed (including many American Catholics) start to think of the Vatican as a bunch of old men in Rome who have some very foolish ideas—ideas that have no relevance to guiding the ethics of their own reproductive efforts.

Whatever good objections come from religious thinking to NST, then, should not be based on fear but on an appreciation of the suffering from genetic disease, of the importance of not harming children, and of the importance of the family. But so far we have not heard much of this kind of thinking.

Improving Humanity

Let us make the optimistic solution that sometime in the not-too-distant future, animal studies have proven that originating a child by NST will likely be very safe. Humans will then have more power to create children who will be free of some common genetic diseases, who may be smarter (more memory, better jokes, more creative), stronger, live longer, more playful, and live happier.

As we know, the fact that we "can" do something does not mean that we "should," and it especially does not mean that we will easily achieve a social consensus that the "can" should be allowed. The question I now want to pose is, when we achieve such results in animals, how will we ever reach a social consensus to allow such results to be used in humans? If we are at ethical point A now, how do we get to ethical point Z?

It is a bit difficult to see how the movement will go, because every time someone proposes that we move to point B, objections are raised that this is immoral "eugenics" and that soon "only perfect babies will be desired." Because of the excesses of past, stupid rhetoric in eugenics, and some terrible past policies, anyone proposing to move society, not to point Z but merely to point B, is labeled as a kook and as a fanatic.

If John Rawls is correct, we clearly owe future generations the best future possible. If we were in the Rawlsian social contract under the veil of ignorance, we would choose to be born as healthy, smart, as strong as possible, and, yes, even as beautiful as possible, with the caveat that there are lots of different standards of beauty (isn't it only honest to admit this?). We would also want people who value knowledge and reason, who care for beings who are sentient, who are courageous, and who are not dependent on drugs that allow them to escape from reality. To the extent that any of these latter qualities turn out to be genetically-related (and none of them may), we would also want such genomes in future populations.

When it comes to non-human animals, we think nothing of trying to match the breed to the needs of the owner. A colleague of mine was thinking about getting a dog. He is a professor of French and returns to his native Paris each summer. He lives in Birmingham in a small apartment. He knows very little about dogs.

I urged him to read books about types of dogs and their degrees of suitability for a life such as his. He was thinking about a large dog, such as a boxer, border collie, or retriever. Each of these breeds is known to be intense and to need lots of exercise. I wanted him to read about breeds of dogs because some breeds, when left alone in apartments, go neurotic, biting and chewing up furniture in fits of loneliness and frustration. Ultimately, after some careful reading, he chose a small dog that he could fit under his seat as he flew back and forth from Paris. The dog now thinks of his apartment as his huge domain.

I do not mean to compare the moral worth of people to that of dogs, but on the other hand, a moral philosophy of future humanity should be realistic. It should never ignore the facts that we are also animals and that we come to be this way partly via the genes we inherit. I have already warned of the dangers of genetic reductionism and genetic fatalism, so I am not advocating that a person is "nothing but" his genes.

On the other hand, it certainly seems true that, just as certain types of dogs would be unhappy in certain kinds of situations, knowledge from the Human Genome Project might eventually allow us to predict—most of the time and for the most people—that certain kinds of people will be unhappy in certain kinds of situations.

Of course, all such generalizations have exceptions and, were we to try to choose people of a certain type for a certain career, e.g., professional athletes choosing to originate the genes by NST of the genome of Tiger Woods or Michael Jordan, there would be many mistakes and false starts. For example, many golden retrievers suffer from an hereditary defect known as hip dysplasia, where the hip socket does not adequately go around the nub of the femur, such that in later life, the femur pops out and cripples the dog, often resulting in the dog having to be put down early while its mind is fine. (Of course, the dysplasia is thought to be due to overbreeding dogs with a bad gene and not having such breeding tempered with sexual reproduction. Myopic selection and control also can be a bad thing.)

So there are always exceptions and dangers to predicting types, but on the other hand, many people love their retrievers and their sunny dispositions around children and adults. Could people be chosen the same way? Would it be so terrible to allow parents to at least aim for a certain type, in the same way that great breeders (such as the monks of New Skete) try to match a breed of dog to the needs of a family?

When it comes to choosing humans by the same general ways that people have chosen dogs and horses for thousands of years, great feelings of unease and distrust arise. One objection is that humans and these other animals are so different that it is demeaning to humans to think of choosing a particular kind of person.

There are two kinds of objections here. The first is an objection to choosing at all, or to making this kind of choice about humans. I think that this first kind of objection has been refuted by this time in this book.

The second objects to applying a kind of choice hitherto used with animals to humans.

This view rests on many misconceptions about animals. For one thing, it rests on a questionable morality that humans and animals should be conceptualized as inhabiting vastly different worlds. In *Created from Animals: The Moral Implications of Darwinism*, James Rachels writes,

> It has always been difficult for humans to think objectively about the nature of non-human animals. One problem, frequently remarked upon, is that we tend to anthropomorphize nature and see animals as too much like ourselves. An opposite but less frequently noticed difficulty is connected with the fact that, even as we try to think objectively about what animals are like, we are burdened with the need to justify our moral relations with them. We kill animals for food; we use them as experimental subjects in laboratories; we exploit them as sources of raw materials such as leather and wool; we keep them as work animals—the list goes on and on. These practices are to our advantage, and we intend to continue them. Thus, when we think about what the animals are like, we are motivated to conceive them in ways that are compatible with treating them in these ways. If animals are conceived as intelligent, sensitive beings, these ways of treating them might seem monstrous.[5]

By conceiving non-human animals as radically different in kind from human animals, we create an artificial barrier in ethics between what can be done to humans and what can be done to other animals. According to Rachels, this makes for an incorrect view of ethics. So the view that whatever can be done to non-human animals cannot be done to human animals, and whatever way we treat human animals is not the way we should treat non-human animals, is wrong on both counts. Instead, Rachels proposes a new way of thinking about all kinds of animals:

> Humans and other animals are not radically different in kind: they are similar in some ways, and different in others, and these differences are often merely matters of degree.

Individuals are to be treated in the same way, unless there is difference between them that justifies a difference in treatment. At the very least, Rachels' view puts pressure on the boundary between what we can do to animals and what we cannot do to humans.

This is also an interesting subject to think about from another direction.

As mentioned before, Ian Wilmut was not really after cloning so much as the ability to transfer genes into a mammal such as Dolly and once a good transfer was achieved, to reproduce that good result indefinitely. This advancement of transfer of a human gene to non-human animals raises many interesting questions.

We know that, despite the views of speciesists, there is a continuum from the primates to man. In some ultimate sense, humans are both nothing more than, and as wonderful as, compassionate monkeys. We know that baboons, gorillas, and dolphins possess the ability to communicate through signs and other sounds. They don't have vocal cords, so they can't talk, but at present we are not sure of the upper limit of their ability to communicate.

Making genetic transfers of human genes to such mammals might not only produce monsters or cyborgs. It might also improve such mammals to the point where they could communicate better and tell us whether they were thinking. (They certainly have feelings.)

That might make life much more difficult for the typical human speciesist who believes that humans are categorically different, and more morally valuable than, baboons. If a genetically-enhanced baby baboon was sent to the best schools, and learned to use a computer to express her thoughts in human voice-sounds, it would be startling to hear her say, "I want to eat" and even more startling, "I entertain you. Watch me jump backwards."

One of the consequences of thinking that animals are not special but humans are is that almost any kind of genetic enhancement may be allowed in any kind of non-human animal. In addition to having cows give better milk, lambs produce better and cheaper drugs, and cattle have better hides, there will be no federal body regulating experiments that may give a baboon or dolphin gene transfers to enhance his ability to communicate. If so, there may come a day when a startled human-person realizes that a baboon is signing, "Hey, I'm a person too!"

So weakening the ethical boundary between non-human and human animals puts pressure to expand the territory in the middle in two directions: to forbid doing to higher non-human animals some of the things we forbid doing to humans, and to allow doing to humans some of the things we think quite sane to do to animals. One item in the latter expansion might be to allow parents to give their babies at birth the best genetic heritage possible.

Alternative Pasts and Social Control

Suppose that in the mid-nineteenth century, when people began to learn that infected water spread cholera, that a chemical had also been discovered that prevented pregnancy when added to the public water system. It was still true that because of cholera, all Western democracies developed public water systems by the end of the century. It would have been easy to have added this chemical to such water systems.

Suppose the chemical acted to prevent ovulation in women. Sex could occur, but no children resulted. To produce children, an antidote was needed and this was dispensed by the State. To get a license to have children, a woman had to prove how many children she had already had (records were kept) and produce evidence that she and her mate were mature and old enough to be satisfactory parents.

Imagine then how counter-intuitive it would seem if today, after more than a century of this practice, it was proposed not to add the chemical to the water any more. It takes no great imagination to conjure the comments that would be made instantly by social conservatives ("We cannot just let anyone have children. What would happen to the human race?")

If such a chemical were ever discovered, I am of course not proposing that it should be added to the water. I am rather suggesting that, if the State used it in the same way as it now proposes to do with originating children by cloning, the problem is that an arbitrary technological innovation would have been used by the State to control and limit human reproduction.

Take another alternative past. Suppose that for centuries in human history only one-fourth of human females could reproduce and that the ability to do so was randomly distributed through some unknown genetic mechanism. Human society has gotten along fine this way, but now something new occurs, viz., a breakthrough in assisted reproduction.

Suddenly it becomes possible that the rest of humanity can reproduce. The catch is that each couple needs the assistance of an infertility clinic to do so: eggs must be carefully primed, sperm must be capacitated and concentrated, several eggs must be implanted to get one successful early fetus, and several attempts at this procedure are usually required to get one pregnancy. Nevertheless, through assistance and tireless efforts, most couples can eventually reproduce.

Imagine the social debate that would occur in this scenario. Traditionalists would decry the new arrangement, pointing out that humanity had gotten along quite fine as things were before, an arrangement obviously created by God. Fatalists would intuit that there was something in the stars that picked certain women to be mothers and would resist any changes. Social conservatives would claim that the very fabric of our society and our fundamental moral values will be ripped apart if just anyone could reproduce. Those enamored of evolution would claim that such assisted reproduction would drag down the human gene pool.

The point is that it is relatively arbitrary what society can control. If social consensus must be achieved before any change in human reproduction is allowed, and if the default position is social control, then human reproductive freedom means very little.

Beneficent Multiples

There are lots of reasons to keep an open mind about whether we should allow couples to deliberately choose to originate certain genotypes by NST. One reason to think about multiply-created children is that the future is open and we do not know what new startling discovery might jolt us into thinking about ourselves in radically new ways. Discovery and contact with intelligent life in other solar systems, who did not resemble humans but resembled, say, very large mollusks, might considerably loosen up our worries about originating humans by NST. If such beings typically reproduced by parthenogenesis, or reproduced by NST half the time, it would soon seem much more natural to us.

Our world changes and so does our fund of facts. As these changes occur, new reasons may appear for why originating multiple versions of the same original genome by NST could be valuable. We should not necessarily close any door that might lead into different rooms in the house of future humanity, especially if it might later turn out to be the only "door" that we needed to enter.

One thought experiment is to imagine what it would be like to have, if male, eight twin brothers or if female, eight twin sisters. Initially, our reaction is one of horror, because science fiction novels have always portrayed the motives and consequences of such arrangements as a bunch of androids having a dumb, group mind. But we shouldn't let the future of human reproduction be so biased by the works of a few past writers who, at the time they wrote, only wanted to titillate us.

Indeed, I am not so sure it would be a bad thing to have eight men existing now who were my twin brothers. We need to think of a future world where such sibling groups are brought about for good motives with good results.

Sets of identical twins that I have known almost always live close to each other. To have one twin move away and be really apart from one's life, twins have told me, would be like the "death of a very important part of me." Like couples who have been married for fifty years and who often die within months of each other, twins need to be physically and emotionally close to each other.

The reason for this is the strong sense of intimacy and support that develops over childhood, adolescence, and adulthood. But imagine such a sense multiplied eight-fold. What a support system I would have! When my life turned bad, there would always be someone available to talk to.

It would be natural to form group projects, such as running a business. People who are more alike get along better, trust each other more. I can't imagine one twin cheating another in business; it would be like cheating yourself.

Over time, some married twins might choose to perpetuate some of the "mother line." This would be especially true for women twins, who could easily carry one mother line over many generations, creating a kind of family dynasty. We can imagine that, instead of Huxley's pejorative names like "alphas" and "betas," such people might proudly identify themselves, such that over time, their family names had the emotional richness that today is conveyed by names such as Spielberg, (Alice) Walker, Einstein, and Roosevelt.

All of which is to say that being a female "Rockefeller" in a line going back centuries through dozens of generations might not be something any person with origins in "cloning" might be ashamed of. Indeed, it is likely that the results would be just the opposite, because if these lines were successful and had members continually passed on, the descendants might easily have great wealth, prestige, and power.

In a show on BBC entitled "Brave New Babies," Jonathan Glover wondered if having 40 adults all looking like Madonna might create oddities in their friendships and relationships with the opposite sex.[6] He interviewed a set of male triplets who were teenagers and all of whom had girlfriends. The latter asserted that each of the triplets was unique and that they were not interchangeable. Glover wondered how it would be if there were 40 twins, but the girls couldn't answer.

But it seems to me that it was kind of a category mistake to ask the question in that way. Although there might be (with his consent, of course) 40 copies of Harrison Ford growing up at various ages, it would rarely be true that there would be 40 little "Harrisons" all growing up together. No one could gestate or raise such a number. To the extent that they were gestated with eggs of different women (different mitochondrial DNA), raised by different women and men, and lived in different areas with different sets of siblings, they would be different persons. (And, hey, the girlfriends of the English male triplets thought so!)

If we adopt a public policy that permits physicians to help couples originate a baby by reproducing the genes of an existing adult, we open many possibilities. One of them is that a rich person may arrange for his wife to gestate a son who originates from his genes and then leave a great deal of money to that son. The father may feel closer to this son than to others and may desire to found a dynasty down through time by creating people with his genes. He may even set up trust funds for future grandchildren and great-children, with the proviso that money can only be inherited if the child originates by cloning and gestating his original genes.

Japan is already famous for ancestor worship, and its culture may help us see how children originated by NST would be fascinated by the life of their founder. Imagine, for instance, reading a diary or journal of one's identical genetic ancestor. Such children might not feel in doing so that they were experiencing a "closed" future but a connection with someone of the same abilities but for whom a past society had offered fewer opportunities. With a little effort, we can imagine worlds where having several twin siblings is neither a tragedy nor a source of anxiety, but a source of pride and even joy.

A Final Prediction

In many past discussions of breakthroughs in reproductive medicine, authors breathlessly predicted that pressing ethical questions would descend on society any day as more radical discoveries were made. Frequently, those new discoveries—the gene for Huntington's, the ability to produce by IVF for most couples who try, a way to really cure a genetic disease in kids through somatic therapy—took a very long time to occur or did not occur at all.

Despite that record, my hunch is that animal studies are going to proceed at a remarkably swift pace. For one thing, such research is not encumbered by the tedious review process of IRBs, the FDA, and the RAC that governs most of human research. Second, animal research is fueled by big business, patent rights, and well-paid individuals who hope to both make a lot of money and to achieve a first. Given these forces, I predict that reliable production of adult mammals by nuclear somatic transfer is only going to take a few years. This includes the reliable production of healthy chimp babies from the same technique.

If so, then the time will come soon when the safety issue diminishes and the only thing stopping trials of NST is, essentially, the expectations argument. When that time comes, I hope we find the courage to try originating a child by nuclear somatic transfer. If we let such a weak argument tie the hands of prospective parents and researchers, we will be doing so only for the flimsiest of reasons, i.e., because of the general unease of traditionalists. To me, the expectations argument would then be insufficient to limit human reproductive liberty. Call me Joe Fletcher's clone.

Notes

1. Robert Nozick, *Socratic Puzzles* (Cambridge, Mass.: Harvard University Press, 1997), 281–2. Jim Rachels brought this passage to my attention in this context.
2. National Bioethics Advisory Commission (NBAC), *Cloning Human Beings: Report and Recommendations of the National Bioethics Advisory Commission*, Rockville, Md., June 1997, 93,
3. NBAC, *Cloning Human Beings*, 32.
4. Ronald M. Green, "At the Vortex of Controversy: Developing Guidelines for Human Embryo Research," *Kennedy Institute of Ethics Journal* 4, no. 4 (December 1994), 348.
5. James Rachels, *Created from Animals: The Moral Implications of Darwinism* (New York: Oxford University Press, 1993), 129.
6. Peter Singer and Deane Wells, *The Reproductive Revolution: New Ways of Making Babies* (New York: Oxford University Press, 1984), 162.

Index

abortion, 87; Old Testament passages, 88; partial-birth, 91
adenosine adaminase deficiency (ADA), 105
agapic Christianity, 165–66
AID (artificial insemination by donor), 6, 99
Alexander the Great, 76
Allen, Woody, 32
Alzheimer's disease, 16, 17
American Association for the Advancement of Science (meeting), 115
amniocentesis, 106
Andre, Judith, 111
Andrews, Lori, 47, 151
Andromeda Strain, 54
animals, 13, 56, 86, 167–70
Annas, George, 2, 42, 52, 54–55, 122–23, 153
Aristotle, 21, 79
artificial hearts, 93
Augustine, 74, 77, 89
Australia, embryo research and, 89
Ayala case, 66

Baby Fae, 86
Backlar, Patricia, 35
bans on embryo research, 85–91; proposed ban on NST, 85

Barnard, Christiaan, 86
Bentham, Jeremy, 59
bioethicists, 3, 27, 33, 35, 67; European, 154
Bladerunner, 40
Boswell, John, 76
boundaries, natural, 124
Boys from Brazil, 29, 42–42, 43, 44–45
brain, why cannot be cloned, 14–15
brain-death, cloning humans and, 48
brains, baby's development and, 50
Brave New World, 56
Brenner, Sidney, 16
Brown, John and Lesley, 4, 47
Brown, Louise, 27–29, 47, 74, 152
Bucher, Glenn, 67
Burcham, Jack, 93

Cahn, Axel, 109
Callahan, Daniel, 27, 132
Cameron, Nigel, 46
Campbell, Keith, 12
capitalism, 163
Caplan, Arthur, 33, 52
Capron, Alex, 35, 135
Carr, Elizabeth, 152
Cassell, Eric, 35
Cavanaugh-O'Keefe, John, 85
Centers for Disease Control (CDC), 155

Gellman, Marc, 124
genes, definition and nature, 14–15
genetic connection, 108–112
genetic determinism. *See* determinism
genetic disease, 86, 103, 113, 132
genetic therapy, 105, 156
George, Peter, 53
Georgetown University, 94–95
Germany, 156
germ-line genetic therapy, 105
Glory Season, 41
Glover, Jonathan, 126, 173
Goldman, Alan, 79
Gould, Stephen Jay, 18–19, 48
Greek, views on sexuality, 76–77
Greenberg, David, 76
Guttmacher Institute, 77

Hafez, E., 68
Hall, Jerry, 11, 31
harm principle, of J. S. Mill, 61, 142
harms, kinds of, 134–35
Hauerwas, Stanley, 119
Hellenism, 76
heterozygote, 16
Hitler, Adolf, 42, 50
Holland (and euthanasia), 70
homonculi, 21
homozygote, 16
Hoppe, Peter, 10
Hornstein, Mark, 33
Hughes, James, 100
Hughes, Mark, 91–95
Human Embryo Research Panel, 42, 90
Human Genome Project, 15–17, 21, 42, 168
humanity (improving), 168
Huntington's disease, 22, 102, 106
Huxley, Aldous, 56

identity, of self, cloning and, 49–51
Illmensee, Karl, 12
imprinting, of genes, 20, 131
informed consent, and cloning, 52, 159
In His Image: The Cloning of a Man, 30–31
institutional review boards (IRBs), 55

Instruction on Respect for Human Life, 73, 75
intracytoplasmic sperm injection (ICSI), 134
intrauterine devices (IUDs), 87
in vitro fertilization (IVF), 4–5; chances of success of, 145; costs of, 107; fraud and, 31; risks and, 25–28; selling eggs and, 160
IRBs. *See* institutional review boards

James, Ron, 38
Jarviks (artificial hearts), 93
Johnson, George, 14
Jones, Howard and Georgeanna, 88
Jonsen, Albert, 34
Judaism, 76–78, 121, 165
Jurassic Park, 32
"Jurassic Park" objection, 105, 123
justice (cloning and), 112–14, 143–44

Kameny, Franklin E., 115
Kant, Immanuel, 45, 47, 59, 109
Kass, Leon, 5, 27, 35, 46, 66, 74, 77–78, 95–96, 108, 125, 137, 139, 144–46
Kennedy, Edward, 93
Kevles, Daniel, 123
Kubrick, Stanley, 53

Lander, Ann, 43
Lappe, Marc, 30
"large offspring syndrome," 12
law, 151–161
Lederberg, Joshua, 68
lesbians, 114–5
Levin, Ira, 42
Levirate marriage, 80
Lewis, C. S., 125
libertarians, 152
liberty. *See* reproductive freedom
literature and cloning, 39–43, 172
Lo, Bernard, 34
Locke, John, 39

Macklin, Ruth, 33, 34, 48
males, bonding and, 111
Marchi, John, 1
Marlow, John, 26

About the Author

. . . . .

Gregory E. Pence is professor of philosophy in the Schools of Medicine and Arts/Humanities at the University of Alabama, Birmingham, where he has taught and written about bioethics for over twenty years. He previously published in the *New York Times, Newsweek,* the *Wall Street Journak,* and the *Journal of the American Medical Association.* He is author of *Classic Cases in Medical Ethics* (2nd ed., McGraw-Hill, 1995), editor of *Classic Works in Medical Ethics* (McGraw-Hill, 1997), and coauthor of *Seven Dilemmas in World Religions* (Paragon House, 1995).